DATE DUE

JE 16 '92			
FE 19 '93			
JE 18 '93			
AP 15 '96			
SE 8 '98			
JE 10 '03			

DEMCO 38-296

URANUS AND NEPTUNE

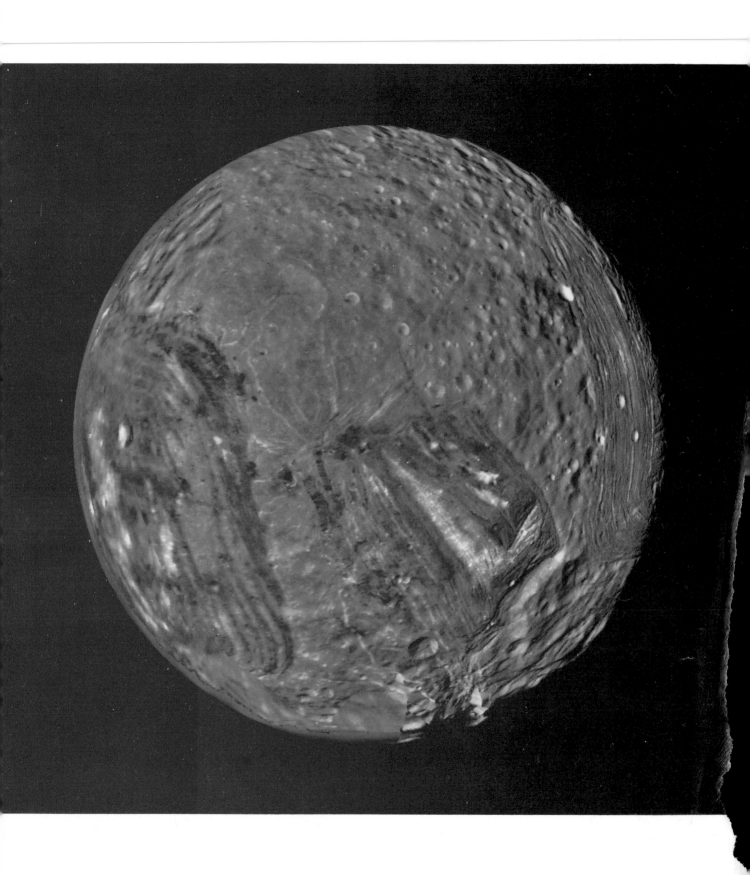

URANUS AND NEPTUNE
THE DISTANT GIANTS

ERIC BURGESS

COLUMBIA UNIVERSITY PRESS
NEW YORK   1988

The Press acknowledges the help of its long-time friend, George V. Cooper, in the publishing of this book.

Library of Congress Cataloging-in-Publication Data

Burgess, Eric.
  Uranus and Neptune.

  Includes index.
  1. Uranus (Planet) 2. Neptune (Planet)
3. Project Voyager. I. Title.
QB681.B87 1988   523.4'7   87-14591
ISBN 0-231-06492-6 (alk. paper)

Columbia University Press
New York   Guildford, Surrey

c 10 9 8 7 6 5 4 3 2

Hardback Columbia University Press editions are Smyth-sewn and printed on permanent and durable acid-free paper

Book design by Ken Venezio

To Staci and Howard Burgess and their daughter
Amanda Marie Grandchild of the Space Age

CONTENTS

PREFACE

October 25, 1986, marked a fiftieth anniversary that passed almost unnoticed by the mass media. On that day in 1936 a group of students from California Institute of Technology, led by Frank Malina and encouraged by Theodore von Karman, fired their first liquid propellant rocket engine in the Arroyo Seco, a rocky dry stream bed above Devil's Gate Dam in the foothills west of Pasadena, California. There, in the shadow of the San Gabriel Mountains, on one peak of which the 100-inch Hale telescope of Mount Wilson Observatory peered deep into space, these young engineers took the first practical step toward realizing their dream of sending machines into space to spearhead a new era of exploration. And from that humble beginning came the Jet Propulsion Laboratory and through it, in turn, came the exploration of all the planets known to the ancients and beyond to Uranus and Neptune, the outer giants. From the Arroyo to Uranus and beyond in less than 50 years was undoubtedly the greatest physical expansion of human frontiers in the history of our species.

Humans have always pushed at new frontiers. But before people move into a new environment they first have to acknowledge that the new environment exists and that it is a place to which they might go. Until thinkers of the Middle Ages had accepted the fact that the Earth might not be flat and that ships might not sail off a limiting edge, there was no point in thinking about exploring beyond the horizon to seek other lands and to circumnavigate the globe. But once the idea of the world as a globe took hold, intrepid explorers began to challenge established limitations and began to penetrate beyond the old horizons.

Today there is often an interim stage, during which the human mind accepts a new frontier, which is then probed by machines before people physically venture into it. This is because beyond the horizons of today there are often alien environments in which it is difficult for humans to survive. Machines are used for the initial explorations, and from their activities the environment of the new frontier can be understood and ways can be found to protect the human explorers so that they can bodily enter the new econiche.

The space immediately surrounding Earth beyond the appreciable atmosphere was first explored and the environment defined by use of sounding rockets and artificial satellites. Later astronauts and cosmonauts went into orbit in space capsules and showed that people could survive in the new environment if suitably protected. Over the last few years Soviet cosmonauts have spent many months in space stations and have operated laboratories in orbit preparatory to interplanetary manned flight.

Apollo astronauts followed Ranger and Surveyor automated spacecraft to the Moon. They showed that people can travel to other worlds and land safely upon them. So far no extended sojourns on the surface of the Moon have taken place, but there are no insurmountable technical difficulties, but only budget restrictions, to prevent us from establishing permanent bases, even human colonies, on the Moon. Such bases can be protected from the harsh lunar environment by totally enclosing them as lunar cities.

Unmanned spacecraft have blazed the way to all the inner planets and to the four outer giants with their fascinating satellites and ring systems. No people have yet traveled on interplanetary flights, although the technical capability is available to send manned missions to Mars. The U.S. has turned away from this achievement, which was a logical follow-on to the Moon landings, and has abandoned this frontier to the Soviets. The Russian space program appears to be concentrating on a manned mission to Mars early in the next century, possibly after exploratory manned missions to the Martian satellites.

For now, however, we rely heavily on unmanned space probes which become extensions of the human senses, extending human touch, sight, hearing, smell, and taste across hundreds of millions of miles and adding additional nonhuman senses that translate their information into terms understandable to humans through the actions of digital computers. By means of these machines we see other planets and their satellites through television eyes, hear through sensitive microphones and seismometers, feel with extension arms and delicate soil scoops, taste and smell with miniaturized chemical laboratories. And often the machines' senses have ranges far beyond human senses, seeing into the infrared and the ultraviolet, hearing sounds that would be inaudible to human ears, and sensing chemical compounds beyond those that can be smelled or tasted.

The accent on inexpensive exploration of the far reaches of the Solar System is by gravitational slingshots which use the gravity and motion of one planet to accelerate a spacecraft to another without the further expenditure of rocket propellants except for maneuvers. Direct flights to more distant planets are too costly in time and too limited in payload capacity to be achieved with launch vehicles of any reasonable size. Advanced propulsion systems could change this, but the development of such systems, started so successfully in the 1960s, has been generally abandoned. However, even with gravity slingshots the entire Solar System can be explored within 25 years or so at a cost less than that spent in the same period to extol on television commercials the capabilities of different brands of over-the-counter pain relievers!

An expansion into the new frontiers of space and the exploration of the outer planets and beyond into interstellar space does not just happen. It is different from a seed germinating on a planet, a baby emerging from the womb, a stream seeking the ocean. It requires dedicated individuals supported by governmental commitment of resources to bring it about. The great leap forward from using the resources of a planet to developing and using the resources of a solar system must be a concerted and definite step, a meticulously planned action. This is because the expansion into space requires such a large effort and capital commitment that it must occupy much national and even planetary thinking and activities. Since machines can, if we wish, now provide humanity with all the necessities of life for all the billions of people on Earth, many of these people can be freed from mental and physical drudgery so as to participate in a creative process of building a future for our species beyond the fragile Earth. Involvement in such a long-term endeavor offers us an opportunity to overcome our current prediliction to use most of our resources to try to protect ourselves from ourselves, like some enormous biological creature whose immune system is running amok.

But for long-term expansion into the Solar System we await the equivalent of Columbus and his Queen Isabella sponsor to develop the Solar System. Undoubtedly such sponsorship will come in the years ahead, even though it is not yet clear who the sponsor will be. And sponsorship need not be on faith alone. It has been suggested that both Columbus and his Queen knew that the Vikings had found a land of many resources and opportunities across the Atlantic Ocean and that the success of Columbus was assured if his transportation system, supplied by his sponsor, did not fail. Today we know also that there are many real worlds in space waiting to be explored and developed, an almost inexhaustible economic commons. We, too, need a sponsor to develop the machines for the exploration because at present the exploration of the outer planets rests with machines; complicated devices of microelectronics and silicon chips possessing fantastic reliability; machines that operate unattended in space for decades, some with the ability to detect faults within themselves and to repair them or switch in replacements for defective elements of their complicated eletromechanical systems. When the automated missions have been completed successfully, machines to carry people into space will also need sponsorship, for there will be no shortage of humans among the children of the space age—barnstormers of the Solar System—clamoring to follow the machines into space when the opportunity arises.

I have been fortunate to have experienced this half century of space explorations, often at first hand. I was involved in the formation of the interplanetary society in England about the time of the Arroyo-Seco tests. Subsequently I witnessed many early rocket tests and their failures, first flights of big missiles and satellites, the testing of great rocket engines and the landing of the Vikings on Mars. I reported the flights of the Mariners and Pioneers pushing in toward Venus and the sun-scorched surface of the innermost planet, Mercury, the mindboggling flybys of mighty Jupiter and majestically ringed Saturn. Secrets of other worlds that were hidden from humanity since its origin have been revealed in these fifty years. And in January 1986 the triumphant flyby of Uranus whetted our appetite for reaching Neptune.

The outer giant planets have always been mysterious worlds, because their great distances limit observations from Earth. Uranus' rotation period was much in doubt (anywhere from 10 to 24 hours). Its satellites were believed to be uninteresting, geologically dead frigid objects. But in the period of a few days more was learned about this strange outer planet by a single spacecraft flyby than in the two centuries of observation since Uranus' discovery in 1781.

Initial planetary encounters are unforgettable experiences for those who are fortunate to witness them. To me the encounter with Uranus was certainly no exception. It was a triumph for those many people whose engineering and other skills brought about this planetary rendezvous at nearly two billion miles from Earth, and continued thereafter to nurse the spacecraft for its ultimate planetary target, Neptune.

Sebastopol, California   Eric Burgess

1

AIM CLOSE TO
THE BULL'S-EYE

The major part of the Sun's planetary system lies beyond the orbit of Mars, in the region referred to as the outer Solar System. It is the domain of the giant outer planets, of very small planetary-type bodies, of comets, and of the boundary between the influence of the Sun and that of the Galaxy. In 1971 the Space Science Board of the National Academy of Sciences stated that an extensive study of the outer Solar System should be a major objective of space science during the next decade. A rare opportunity to visit several giant planets was presented by an unusual configuration of these bodies; furthermore, missions to the outer Solar System would permit the exploration of the boundary between the Solar System and interstellar space.

To meet the broad challenge of answering questions about the origin and evolution of planetary systems and interactions of stars such as our Sun with the Galaxy, the outer Solar System needed to be explored and meticulously studied because the state of human knowledge about this vast and important region was fragmentary and limited. As a result, four spacecraft were allocated by the National Aeronautics and Space Administration (NASA) to explore the outer Solar System, Pioneers 10 and 11, and Voyagers 1 and 2.

After flying by Jupiter, Pioneer 10 headed out of the Solar System, Pioneer 11 and Voyager 1 flew by both Jupiter and Saturn before heading out of the Solar System, and Voyager 2—after flying by Jupiter and Saturn—continued on to Uranus and Neptune.

Uranus is the seventh planet outward from the Sun which it orbits every 84 years at an average distance of 1.783 billion miles (2.87 billion km), which is 19.2 times Earth's average distance from the Sun. Miranda is the smallest of the satellites of Uranus observed from Earth. Its diameter is about 300 miles

(500 km). Miranda was the most recently discovered of the five known Uranian satellites—by Gerard C. Kuiper in 1948. Named after the daughter of Prospero, a character in Shakespeare's *The Tempest*, Miranda orbits Uranus in a period of 1.4 days at a distance of approximately 80,660 miles (129,800 km). When observed from Earth, even in the best telescopes, it appears only as a very faint, star-like object.

Because of the great distance of Uranus, the size of Miranda and its position at any given time cannot be precisely determined by observations from Earth. Yet Voyager 2, a spacecraft carrying a complex battery of science instruments to explore Uranus, was targeted to fly between Miranda and the planet at 45,000 mph (72,500 kph) on January 24, 1986. The goal was to obtain detailed close-up pictures of this tiny frozen world, as well as of Uranus and its other satellites, to find out more about present conditions and the evolution of worlds of the outer Solar System.

On January 23, 1986, as Voyager 2 sped toward the Uranian system, images of Miranda transmitted to Earth from the spacecraft were revealing the satellite's surface for the first time (figure 1.1). They showed subtle albedo markings—light and dark areas—and a peculiar bright chevron-shaped feature, quite unlike anything seen elsewhere in the Solar System.

By imaging the planet Uranus and its satellites against starry backgrounds over several days before closest approach of the spacecraft to Uranus, navigators and mission controllers had been able to update instructions stored in the electronic memory of the spacecraft to aim its movable platform on which cameras were mounted. The task was to direct a high resolution (telephoto) camera accurately so that eight overlapping frames would cover the whole visible disc of Miranda as the spacecraft raced past it some 17,400 miles (28,000 km) above the uncharted surface of the satellite. The camera also had to be moved to compensate for the relative motion of spacecraft and satellite during each of the long exposures necessary to obtain images of the dark and faintly-illuminated surface.

On January 25, 1986, at 1:00 P.M., the image monitors at the Jet Propulsion Laboratory, Pasadena, California, from where NASA's Voyager spacecraft was controlled, started to display the close-up images of Miranda obtained the previous day by Voyager. Observers gasped at a great valley biting several miles into the limb of Miranda (figure 1.2). An hour later even more astounding images were displayed of the bizarre surface of this tiny world. There were linear markings as well as heavily cratered terrain and unique features never before seen on worlds of the Solar System. The chevron marking was revealed as a complex system of bright and dark linear faults—long fractures showing that displacements of blocks of surface had taken place. A bewildering variety of other fractures, of grooves and of craters, appeared on terrain with different shadings, some very dark, others relatively bright (figure 1.3).

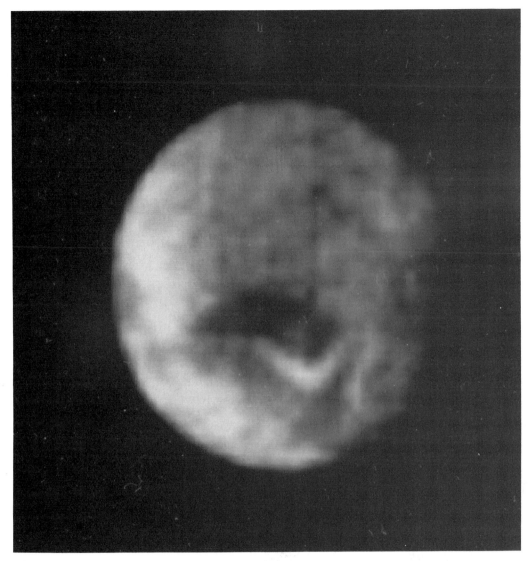

FIGURE 1.1

The first views of Miranda at a distance of 860,000 miles (1,380,000 km) showed a sinuous bright marking crossing a dark area near the pole of the satellite. On the left was an area of brighter albedo. On the other side was a larger area of medium albedo. Much finer detail tantalized at the limits of resolution with unusual-looking projections from the terminator and irregularity on the limb. (Photo. NASA/JPL)

Miranda displayed three distinct surfaces—cratered, grooved, and jumbled—as though the satellite had been taken apart and put together again haphazardly. The ancient cratered terrain consists of rolling, subdued hills and degraded craters of medium size. The grooved terrain displays linear valleys and ridges, and the complex terrain of intersecting curved ridges and troughs has abrupt

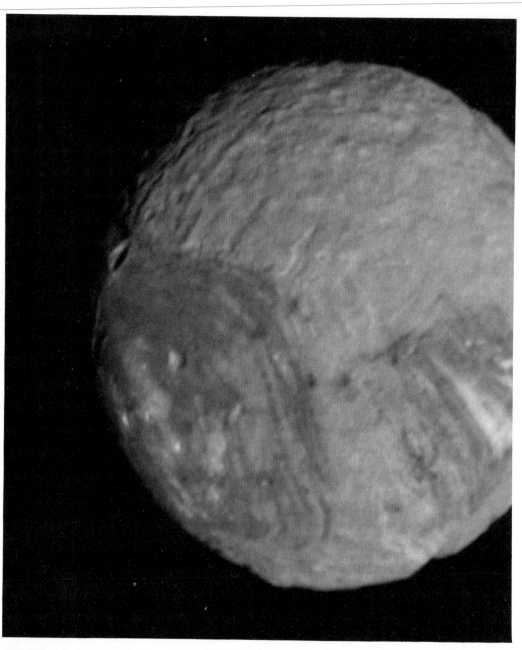

FIGURE 1.2

At a distance of 91,000 miles (147,000 km) a colored mosaic shows a grey surface pockmarked by craters but with many unusual features of grooves and ridges. The sinuous bright marking is revealed as a chevron shape, while the area of dark albedo appears as a banded rectangle with rounded corners. The central part of the rectangle is spotted with bright splotches. On the limb a high mountain is silhouetted against the darkness of space, and not far from it is an enormous chasm. The surface is virtually colorless. (Photo. NASA/JPL)

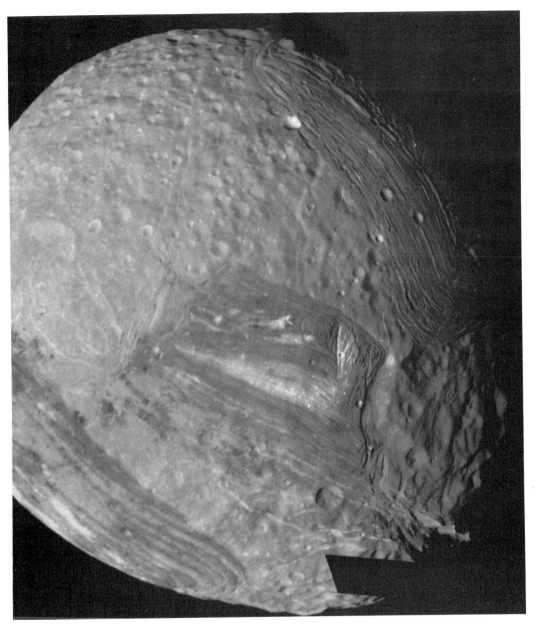

FIGURE 1.3
Imaged at about 20,000 miles (32,000 km) from Miranda the bright chevron feature is revealed as complex parallel ridges forming part of a trapezoidal complex with bright and dark albedos. The rectangular feature is seen as another involved complex of ridges and troughs with a similar ovoid feature on the other side of the chevron. Even the cratered areas are seen to be fractured by many faults. Ridges and valleys are cut off against the boundary of the adjoining province. There are compressional faults and extensional faults as some parts of the surface appear to have been pushed together while other parts have been pulled apart. Many of the faults exhibit large scarps or cliffs ranging from 0.3 to 3 miles (0.5 to 5 km) in height. They are much higher than the walls of Earth's Grand Canyon and more like the walls of the big canyons on Mars. (Photo. NASA/JPL)

FIGURE 1.4

One of the closest images of part of Miranda was obtained at a distance of 22,000 miles (36,000 km). This picture shows objects about 2600 feet (800 meters) across. Two distinct types of terrain are the rugged, high-elevation area on the right and the lower, striated terrain on the left. Note, however, that there are striations on the high plateau area. A spectacular feature at lower right is the sunlit cliff faces where the valley floor has dropped down. This same valley reaches across the sunlit face of the satellite to the deep depression seen on the far limb. (Photo. NASA/JPL)

boundaries at linear grooved terrain. Ridges and valleys of one province were seen cut off against the boundary of an adjoining one. Some faults appeared to have been caused by compression where surfaces had been pushed together. Others appeared to have been caused by extension as the crust had been pulled apart. Miranda was apparently a world with a very complex geological history.

"Just mind-boggling," commented a planetary scientist, that this small body has such a variety of terrains.

Perhaps the most bizarre feature was a series of sunlit flat plates (figure 1.4) seen projecting beyond the terminator, the boundary between day and night on the satellite. When this feature was first seen on the monitor screens its character was debated hotly. Later these bright areas proved to be enormous grooved cliffs rising almost 3 miles vertically from a satellite-encircling valley originally seen as the cutout from the limb in an image displayed earlier. The valley walls are several times higher than those of Earth's Grand Canyon. Many saw-toothed terraces fringe the valley. The cliffs drop from a rugged high terrain with degraded old craters to a striated valley floor (figure 1.5).

FIGURE 1.5
A blowup of the scarps reveals vertical grooves caused when the block of terrain slipped downward. These cliffs are 3 miles (5 km) high, far exceeding any near vertical cliffs on Earth. On our planet the surface gravity is too high for such high cliffs to survive, the material would collapse into and fill the valley. Many terraces are shown elsewhere on the valley walls with smaller lines of cliffs revealing multiple faulting. Note also the shadows reaching across the floor of the valley toward the bases of the very high cliffs.

Elsewhere, two equally unusual land forms consisting of grooved terrain are seen arranged in patterns, each of which is like an enormous race track.

It was several days before these strange features could be evaluated and explained, at least in part.

The close encounter with Miranda was an unforgettable highlight of the exploration of the Uranian system. But there were many other surprises and exciting incidents as the seventh planet was explored for the first time.

The successful Voyager exploration of the Uranian system was the culmination of years of activity. Nearly nine years earlier, Voyager 2 had begun its long journey into space atop a Titan III-E/Centaur rocket. It was launched from the complex at Kennedy Space Center, Florida (figure 1.6) at 10:29 A.M. EDT, August 20, 1977, only a few minutes after the opening of the launch window— the few weeks during which the relative positions of Earth and planets were appropriate for the mission to the outer Solar System.

Could we have seen the stars and planets in the sky during the daytime, we would have identified two targets for the spacecraft; giant Jupiter in the constellation Gemini setting in the southwest, and Saturn close to the Sun in Leo and rising in the southeast. The other two targets, Uranus and Neptune, were below the horizon at time of launch.

If the spacecraft had been programmed to fly directly to Uranus it would have taken thirty years with the rocket launch power available. But by using the gravity and orbital motion of Jupiter and later of Saturn, Voyager was targeted to reach Uranus in 1986, the mission was shortened from thirty years to nine. This was still a formidable task requiring high reliability on the part of the spacecraft and its millions of individual components, and dedication on the part of ground crews and scientists to concentrate their careers for over a decade on a single mission.

Uranus is an oddball planet. It rotates on its side, so that at times it points one pole toward Earth and Sun and 42 years later the other pole (figure 1.7). Equatorial regions face sunward during part of the periods in between. Voyager headed toward Uranus when the planet's south pole was pointing toward the Sun and Earth. The International Astronomical Union defines the south pole as the one which is below (i.e., south of) the orbital plane of the planet.

Uranus' five known satellites move in orbits close to the equatorial plane of the planet. So, as viewed from Earth at the time of the mission, the satellite orbits looked like a target with the planet at the bull's-eye (figure 1.8). The Voyager was aimed toward a close approach to that bull's-eye, passing within 66,500 miles (107,000 km) of the cloud tops of the planet and just inside the orbit of the innermost satellite, Miranda.

When the Voyager spacecraft flew by Jupiter and Saturn it had the great advantage of not being the first to do so. Two Pioneer spacecraft, Pioneers 10 and 11, and another Voyager spacecraft, Voyager 1, had blazed the trail to the giant planets and established the nature of their environments. At Uranus, however, there was no precursor mission: Voyager 2 became the pioneer, and because of the great distance of Uranus, about four times as far as Jupiter and twice as far as Saturn, there were many uncertainties and unknowns about

FIGURE 1.6
At 10:29 A.M. EDT, August 20, 1977, Voyager 2 was launched on a voyage that would carry it to four planets, Jupiter, Saturn, Uranus, and Neptune. (Photo. NASA)

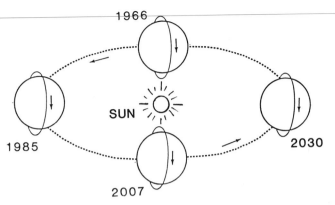

FIGURE 1.7
Because Uranus rotates on an axis inclined 98 degrees to the plane of the planet's orbit around the Sun, different aspects of the planet are presented to the Earth during its 84-year orbit. At the time of Voyager's encounter with Uranus the south pole of the planet faced Earth.

Uranus. Most important of these from a mission planning standpoint was the precise location of the planet and particularly of tiny Miranda.

Operating a spacecraft at the distance of Uranus also presented new challenges. The round trip communication time is 5.5 hours compared with 92 minutes for Jupiter and just less than 3 hours for Saturn. At Uranus the Sun

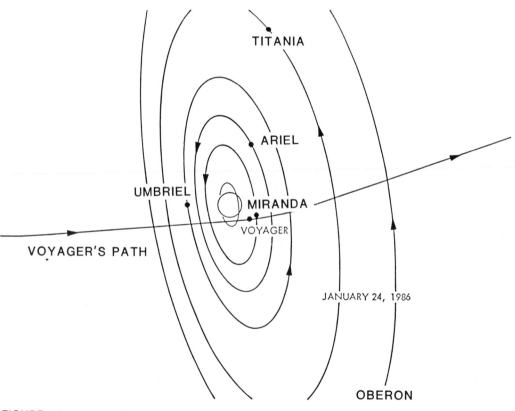

FIGURE 1.8
The orbits of the five major satellites appear like a giant archery target to the approaching Voyager spacecraft in January 1986.

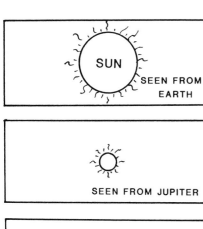

FIGURE 1.9
The size of the Sun seen from Uranus and Neptune is indicative of the low light levels at these distant planets, their faintness in the skies of Earth, and the difficulties in imaging the planets and their satellites.

shines more like a very brilliant star than the blinding disk of the Sun we are used to on Earth (figure 1.9). The intensity of sunlight at the Uranian system is 1/400 that at Earth, about 1/15 that at Jupiter and 1/4 that at Saturn, so that longer camera exposures are required to obtain images of the planet and its satellites. Because the trajectory passed almost vertically through the plane of the satellite orbits, the closest encounter with all satellites occurred at almost the same time. As a consequence rapid retargeting of the camera system was required to obtain images of each satellite in the short time available—less than six hours to obtain close-up images of the planet, its rings, and its satellites. This was very different from the Jovian and Saturnian encounters when several days were available for obtaining detailed pictures. At Uranus imaging required rapid slewing of a camera mount that had encountered severe problems during the earlier encounter with Saturn.

To retrieve images from the spacecraft the data had to be compressed within

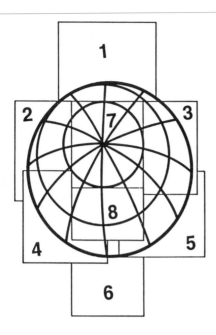

FIGURE 1.10
At the closest approach to Miranda, timing and pointing of the imaging instrument had to be perfect to make sure that the eight imaging frames would cover the entire surface of the small satellite.

the spacecraft before transmission. This required that the onboard equipment be reconfigured during the cruise between Saturn and Uranus to use a different encoding system for the data.

To obtain sharp images of the satellites during the 45,000 mph (72,500 kph) flyby of Uranus, image motion compensation had to be used. This is similar in concept to panning a camera to photograph a rapidly moving vehicle. Again this required use of the faulty platform slewing mechanism. It also required that the precise position of the satellites, particularly of Miranda, should be known. Otherwise the image might not include the surface of the satellite and the pictures returned to Earth would show only the blackness of nearby space.

During the approach of Voyager to Uranus, the cameras had been used to pinpoint stars relative to the spacecraft and the Uranian system and thus to navigate optically through the system. Each day the accuracy of positioning was improved so that commands to the spacecraft could be updated to take account of the more accurate knowledge of the positions of the Uranian satellites, particularly of Miranda, relative to the spacecraft. The Miranda ephemeris (the position of Miranda at any given time) was most critical because the spacecraft was intended to fly 17,400 miles (28,000 km) from that satellite. Errors of position had to be kept as small as possible so that the planned eight image frames at very high resolution would cover all the surface of the satellite visible from the spacecraft (figure 1.10).

The project's objective was to navigate Voyager to within 62 miles (100 km) of its target—comparable to sinking a golf putt at a distance of 1500 miles! The five known satellites (figure 1.11) are relatively small compared with

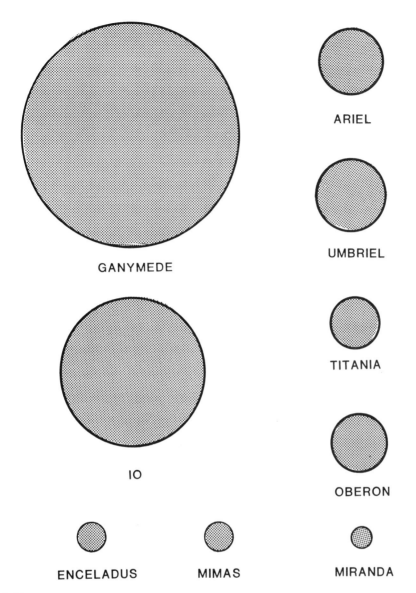

GANYMEDE

ARIEL

UMBRIEL

IO

TITANIA

OBERON

ENCELADUS

MIMAS

MIRANDA

FIGURE 1.11
Relative sizes of the five known satellites of Uranus compared with some satellites of Jupiter and Saturn.

those of Jupiter and Saturn. Titania and Oberon, each less than half the diameter of our Moon, were discovered by William Herschel six years after his discovery of Uranus. Ariel and Umbriel are about one-third the size of Earth's satellite, and they were discovered in 1851 by William Lassell. Tiny Miranda, less than half the diameter of Umbriel, was not discovered until 1958, when Gerard P. Kuiper found it.

An unusual feature appeared to be that unlike the major satellites of Jupiter

FIGURE 1.12
NASA's Airborne Infrared Observatory, a Lockheed C-141 jet transport equipped with a 36 inch (91.5 cm) telescope was used to observe an occultation of a star by Uranus which led to the discovery of the planet's rings in 1977. (Photo NASA/Ames)

and Saturn, the densities of the Uranian satellites increased with distance from the planet. However, observations from Earth are difficult, and densities were very uncertain as Voyager rushed toward the planet. Generally, however, the Uranian satellites were known to be denser and there was speculation that they were darker than would be expected of small frozen worlds, even though their surfaces were known to be icy.

About the time of launching of Voyager, an important discovery was made from observing an occultation of a star by Uranus. Observations were made by Robert Mills in Australia and James Eliott in a high-flying NASA aircraft as his observatory platform (figure 1.12). It was discovered that the light from the star blinked on and off several times before the time when the star was scheduled to be occulted by the planet, and afterwards the on and off sequence was repeated in reverse. These observations were interpreted correctly as being

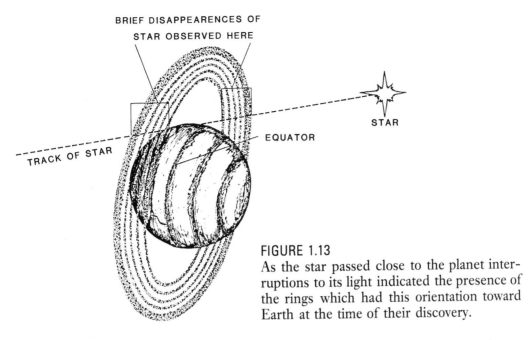

BRIEF DISAPPEARENCES OF
STAR OBSERVED HERE

STAR

EQUATOR

TRACK OF STAR

FIGURE 1.13
As the star passed close to the planet inter-
ruptions to its light indicated the presence of
the rings which had this orientation toward
Earth at the time of their discovery.

indicative of the presence of a system of dark rings around Uranus (figure
1.13), a ring system that could not be observed directly from Earth by light
reflected from it. Subsequent ground-based observations confirmed that this
was so, and in 1979 the Voyager spacecraft also found that Jupiter had a
tenuous ring system that could not be seen from Earth. It began to look as
though ring systems were common features of the large outer planets.

Observing the Uranian ring system from Voyager became an important aspect of the mission to the outer gas giants, especially as there were also some
observations indicating that Neptune too has a ring system.

In April 1985, as Voyager continued toward Uranus, Bradford A. Smith,
working with a charge-coupled device attached to a 100-inch (2.5-m) telescope
at Las Campanas, Chile, succeeded in obtaining an image of the Uranian
system from Earth. The charge-coupled device is an advanced electronic detector which amplifies light to produce images of very faint objects. This image
showed the planet and its known satellites and the ring system. Although the
individual rings were not resolved, the image confirmed that the rings were,
indeed, present at the expected distance from Uranus, and that they are very
dark.

The darkness of a surface is measured by *albedo*—the percentage of incident
light reflected back from that surface. The Earth's albedo is 39%, that of fresh
snow is 78%, and that of carbon black is 3%. The albedo of the Uranium
rings is 1.5–2%. Scientists speculated that the ring material was either organic
material rich in carbon or frozen methane with its surface darkened by energetic particles trapped in yet-to-be-confirmed radiation belts.

The prospect of observing these rings at close quarters was intriguing. Would they have structure like those of Saturn? Would they have shepherding satellites as discovered at Saturn? Would they reveal thousands of ringlets when back-illuminated as Voyager flew beyond Uranus? Would there by symmetry? Would some rings be braided?

By July 1985 Voyager was obtaining images showing Uranus as a disk, and star-like images of the four larger satellites. Miranda, because it is so small and faint, could not be detected on these images. By November 4 of that year, the images of Uranus had better resolution than can be obtained from Earth. But no details could be discerned on the greenish-blue disk of the planet. The planetary disk showed no clouds or zonal bands of the type seen on Jupiter and Saturn. Series of images like those taken at Jupiter and Saturn to determine rotation rates from the movement of atmospheric features were of no use to determine the rotation rate of Uranus.

Toward the end of November, however, six images of the Uranian system were combined by computer processing to show the outermost and brightest ring; the epsilon ring. The image showed the broadening of the ring on one side of the planet as had been predicted from occultation measurements.

It was not until a month later, December 27, that another series of five images, combined into one by computer processing, showed the first clear evidence of latitudinal atmospheric bands on Uranus (figure 1.14). They were, nevertheless, very subdued compared with the zonal bands of Saturn (figure 1.15) which, in turn, were subdued compared with those of Jupiter (figure 1.16). A dark hood over the pole of rotation of Uranus was surrounded by bands of differing levels of brightness, but still with no evidence of cloud patterns.

Almost another month passed before cloud features were identified that could be used to confirm the period of rotation of the planet.

Meanwhile the approach to Uranus had produced other mysteries. Jupiter and Saturn, rapidly rotating gas giant planets, had intense magnetic fields, and Uranus also was expected to have such a field. Scientists had come to expect that any planet possessing a magnetic field would be a source of radio emissions, produced by charged particles trapped in the magnetic field. Jupiter's radio signals were detected by Voyager as soon as the spacecraft left Earth. Saturn's emissions were detected soon after the spacecraft passed Jupiter. At the beginning of December 1985, when Voyager was about 43 million miles (70 million km) from Uranus, the spacecraft's instruments had still not detected any radio emissions from the planet.

Observations by an Earth satellite of ultraviolet emissions from Uranus had been interpreted as indicating the presence of aurorae which, in turn, required that the planet should have a strong magnetic field. Aurorae are generated when charged particles, predominantly electrons and protons, travel along

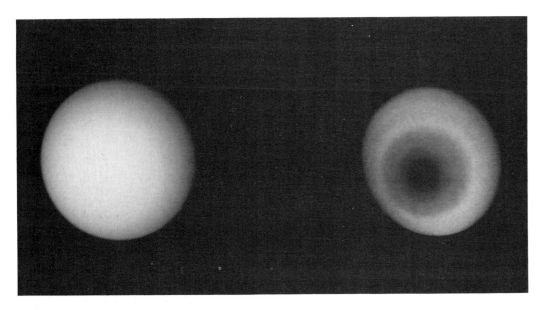

FIGURE 1.14

These two pictures of Uranus were compiled from images recorded on January 10, 1986. The left-hand image is processed to show Uranus as it would appear to the human eye from a distance of 11 million miles. It shows no detail on the blue-green disk. The image is a composite of blue, green, and orange filtered images. The picture on the right uses false color and contrast enhancement to bring out subtle details on the planet. It reveals a dark polar hood surrounded by a series of progressively lighter concentric bands. The dark polar hood may result from smog-like aerosols produced by the action of sunlight on methane. Solar ultraviolet may be producing acetylene, which is reacted upon further to make reddish-brown particles. The banded zones of Uranus are extremely faint compared with those of Jupiter and Saturn. (Photo. NASA/JPL)

magnetic field lines toward the north and south magnetic poles and interact with atoms of the upper atmosphere. The question was, why were no radio emissions detected if there were aurorae on Uranus? Perhaps the ultraviolet emissions were not from aurorae but from airglow; atoms of the atmosphere emitting energy absorbed from sunlight. Perhaps Uranus did not have a magnetic field after all, or only a very weak one. But if so, this would make it a very odd planet.

The spacecraft continued its approach to Uranus. In mid-January 1986, at a distance of about 2 million miles (3.2 million km), no radio emissions were yet detected. But finally, on January 24, only 10 ½ hours before closest approach to Uranus, a bow shock was observed at a distance of 10,500 mi (17,000 km) from the planet. This confirmed the presence of a magnetosphere that was later charted. A bow shock is created when the high velocity stream of charged particles of the solar wind is abruptly slowed down on encountering the mag-

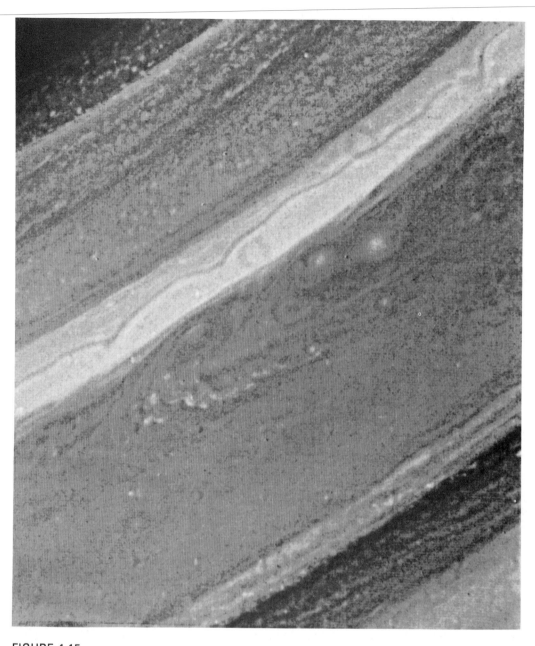

FIGURE 1.15
Saturn, by contrast to Uranus, has more definite zonal bands. This is believed to be because clouds in the atmosphere of Saturn are at a higher level than those in the atmosphere of Uranus. (Photo. NASA/JPL)

FIGURE 1.16
At Jupiter, clouds are even higher and the zones become extremely pronounced and show more intricate details than are seen on Saturn. (Photo. NASA/JPL)

netic field of a planet. The magnetosphere is the region where the effects of the magnetic field of the planet dominate over those of the magnetic field of the solar wind. In the magnetosphere energetic particles are trapped by the magnetic field of the planet. The magnetosphere is shaped somewhat like a windsock with a blunt head facing the Sun and a tail streaming away in the opposite direction. Voyager had finally answered the important question about Uranus by showing that the planet does have a large and unusual magnetosphere, and a strong but peculiarly aligned magnetic field. The magnetic axis of Uranus was found to be offset 60 degrees from its rotational axis, many times more than the offset for any other planet's magnetic field. Earth's magnetic axis is currently offset by about 12 degrees.

By January 20 images of all the satellites had been returned to Earth and showed a great variety of surfaces, all different from one another (figure 1.17). Umbriel is the darkest, reflecting only about 12% of incident sunlight. Oberon and Titania reflected about 20%, Ariel and Miranda about 30%. Ariel showed the greatest contrast of surface features with its brightest area reflecting 45% of incident sunlight. In color images the satellites were of very similar color—a nondescript dirty grey. The Voyager imaging team had also added a number of new satellites to the Uranian entourage—two shepherding satellites and eight other small bodies, most only about 30 miles (50 km) in diameter.

In revealing its secrets Uranus was turning out to be a strange world and to be part of an even stranger system. Scientists were obtaining thousands of times more information in a few days than had been obtained in the two centuries of Earth-based observations since the discovery of the planet by William Herschel in 1781. This had all been made possible through the longevity and versatility of a single spacecraft originally designed to explore Jupiter and Saturn, coupled with the ingenuity of hundreds of scientists, engineers, and support people who directed the spacecraft from here on Earth.

The Voyager mission had received Congressional approval in 1972. Recommendations for the exploration of the outer Solar System had been given in a report of the Space Science Board of the National Academy of Sciences in June 1969. The study of the outer Solar System, said the report, offers a rich field for the discovery of new phenomena, for the development of new concepts, and for definitive solutions to long-recognized problems in physics, astronomy, and chemistry. The report also pointed out that during the 1970s exceptionally favorable astronomical opportunities would occur for multiplanet missions by a single spacecraft, which would not occur again for nearly 180 years. It was emphasized that missions to the outer planets necessitated long-range planning starting many years in advance of the launch date.

Many scientists stressed the significance of the planetary opportunity presenting itself at a time when the technical capability to exploit it was also available.

A major problem faced by all theories about the origin of the Solar System and its planets is that of determining the chemical composition of the nebula out of which the Sun and its planets were formed. The outer Solar System, because it contains most of the material in the Solar System, can provide a

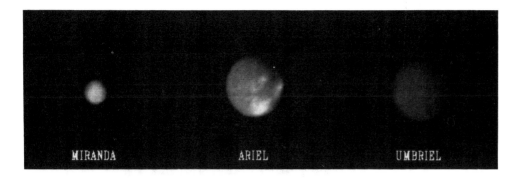

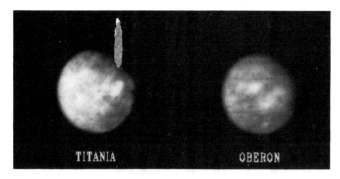

FIGURE 1.17

A family portrait of the large satellites of Uranus showing the relative sizes and reflectances of the satellites. From left, in order of increasing distance from Uranus, the satellites are Miranda, Ariel, Umbriel, Titania, and Oberon. Oberon and Titania, the largest satellites, are about half the diameter of Earth's Moon. Miranda, the smallest of the Uranian satellites, is about one quarter the diameter of our Moon. Even in these images, taken by the imaging system carried aboard the Voyager spacecraft at distances ranging from 3.1 to 3.8 million miles (5.0 to 6.1 million km), the Uranian satellites appear as very individual worlds. Ariel shows the greatest amount of contrast on its surface. Umbriel is the darkest and Miranda the brightest of the satellites, although Titania looks the brightest on this montage. (Photo. NASA/JPL)

treasure trove of materials, some of which have not been changed for billions of years or have not been lost during planetary evolution as was the case with the inner planets and the Earth. On Earth, for example, primordial materials have been extensively reprocessed during the evolution of our planet. Almost all the mass and angular momentum of the Solar System outside of the Sun, and a great diversity of physical properties, are associated with the giant planets, Jupiter, Saturn, Uranus, and Neptune and their attendant satellite worlds.

In 1969 and 1971 the Space Science Board issued further reports again emphasizing the importance of exploring the outer planets and of allocating a greater fraction of NASA's budget to such exploration. An extensive study of the outer Solar System was regarded as being of great importance to space science during the following decade. The report endorsed the TOPS Grand Tour concept as an impressive opportunity to use existing technology to explore the outer planets. TOPS was an acronym for Thermoelectric Outer Planet Spacecraft; thermoelectric energy being needed to generate electrical power for a spacecraft at great distances from the Sun because solar radiation, decreasing as the square of the distance, would be impractical beyond the orbit of Mars— the area of solar cells needed would be enormous. Further, the Space Science Board recommended that the Grand Tour be augmented by probes to enter and sample the atmospheres of the giant outer planets. The Grand Tour should also be supplemented by less expensive Pioneer-type spacecraft to explore individual planets in greater detail.

The TOPS spacecraft for the Grand Tour was proposed as a spacecraft stabilized on three axes, with at least a 10-year lifetime of operations in space at great distances from the Sun. It had to possess the ability to repair itself during its journey and the capability of being reconfigured during its flight from one planet to another. The cost to develop TOPS was estimated at $440 million; the program itself would cost around $800 million.

In April 1971 NASA selected a team of 108 scientists to participate in the definition of the science objectives of the TOPS missions to the outer planets. Funds to begin developing the spacecraft were requested of Congress in the budget submitted by President Richard Nixon for Fiscal Year 1972. The name Grand Tour was first associated with a four-planet mission—Jupiter, Saturn, Uranus and Neptune (JSUN mission). A variety of Grand Tour missions were possible, but two distinct missions were later selected as preferable because they allowed flybys of all five outer planets instead of only four, which would be possible with a single spacecraft. These two missions were a 1977 launch to fly by Jupiter, Saturn, and Pluto, (JSP mission) and a 1979 launch to fly by Jupiter, Uranus, and Neptune (JUN mission). The JSP mission would encounter Pluto near its perihelion, which would provide the best lighting conditions in 248 years for obtaining images of the small, dimly illuminated, outermost known planet.

John E. Naugle, NASA's Associate Administrator for Space Science, announced, on February 24, 1972, NASA's decision to retreat from the Grand Tour of the outer Solar System and to substitute two flybys of Jupiter and Saturn. The new plan, which was forced on NASA by budgetary restrictions and congressional recommendation, would use spacecraft based on the Mariners that the Jet Propulsion Laboratory had developed to explore Mercury, Venus, and Mars. The project was, indeed, named initially Mariner Jupiter-

Saturn. A 1975 report from the Space Science Board of the National Research Council recommended that NASA should undertake a Mariner Jupiter-Uranus mission and also recommended that there should be an immediate re-evaluation of strategies for missions to explore the outer Solar System during the following decade.

A Mariner Jupiter-Uranus mission was later scheduled for 1979, but ran into a budget squeeze that put the $177 million project in jeopardy. It was reported in December 1975 that NASA might use the Mariner Jupiter-Saturn spacecraft, scheduled for a 1977 launch, as a means to permit a flight to Uranus as an extension of the mission after a successful encounter of the spacecraft with Saturn.

Why this interest in the outer Solar System?

The big outer planets are radically different from the small inner planets. While the inner planets are solid bodies of roughly the same size and are all of relatively high density, the very much larger outer planets are of low density because of their enormous atmospheres. The exception is distant Pluto, which is only about the size of Mercury. Pluto's orbit is inclined 17 degrees to the general plane of the Solar System and is extremely eccentric so that at times Pluto is outside the orbit of Neptune while at other times it is within the orbit. Pluto, as seen from Earth, can move far south and north of the ecliptic path followed by all the other planets.

Exploration of the outer planets presented a unique opportunity for the United States because it required two capabilities in which American science and technology excelled during the 1970s—the ability to navigate and command spacecraft over immense distances, and the ability to return information from those distances. No other nation came near to matching those capabilities then. Yet funding for such exploration was reaching new lows in the American budget following the economic disaster of the Vietnam conflict.

It is often difficult to see the importance of exploring the outer planets when confronted with the problems besetting humankind here on Earth. Unfortunately, none of the familiar human problems—inflation, poverty, energy shortages, pollution, crime, warfare—have ever been solved by direct attack, mainly because their solution often affects special interests of one kind or another and requires a redistribution of wealth or of territory. As Margaret Mead once said: "The argument that if we did not spend money on battleships we would spend it on education is fallacious. There is not the slightest proof that if we had not been working on space we would have done anything of greater human value."

In the past, many of the problems of poverty and unemployment have been eased by major wars, but at a terrible general sacrifice of personal property and safety. The space age by contrast, in the late 1950s and early 1960s, stimulated the nation's economy without such sacrifice. It developed new technologies, and new products, and even encouraged the building of new cities. Unfortunately all this slowed down with the advent of the Southeast Asia conflict (1963–1973) which spawned an increasing anti-technology backlash and caused an enormous drain upon the nation's resources.

Many times exploration has substituted for war and satisfied mankind's inner need for expansion. But even more important, exploration changes outlooks. Through it people find relief from competition in a restrictive environment and they also find changes in value systems. But exploration has always been governed by politics—Columbus and Isabella, the Louisiana Purchase, the Westward expansion, the Alaskan ice box. And politics is basically compromise: the reconciliation of different viewpoints, different demands, and different priorities—in hope of providing for the common good.

While dreamers may dream of building empires, of conquering worlds, of exploring the Solar System, unless they marshal resources to provide capital and labor, their dreams can never become real.

For as long as people have peered at the outer planets of the Solar System through telescopes, these great planets beyond Mars have fascinated and intrigued them. Massive and colorful, and rich with satellites that rival the inner planets in size, these planets exhibit unique features and strange characteristics; Saturn's wafer thin rings, Jupiter's internal heat like a gestating star, Iapetus, a Saturnian satellite with one hemisphere six times as bright as the other, Uranus spinning in its side with coal-black threads of rings, and Neptune with a huge satellite going in the wrong direction! Beyond frigid Pluto, on the brink of the interstellar abyss, the solar wind tangles with interstellar particles and stops its outward rush from the Sun at a hypothetical region called the heliopause. And ever further out there is believed to be a region of comet spawning which periodically releases these messengers from the outer darkness to stretch glowing tails of plasma and dust across the inner planetary orbits as they swoop around the Sun in their perihelion passages.

The elemental composition of the outer planets may be rich in clues to the origin of the Earth and of humankind. Laboratory experiments led to speculations that prebiological activity, reminiscent of the primordial life-spawning activities on Earth, may now be taking place on the worlds and large satellites of the outer Solar System.

Beyond these strange planets and their retinues of smaller satellite worlds, the stars beckon from the unexplored void of interstellar space; billions of stars around which may be circling planets inhabited by creatures with intelligence equal to or greater than our own. Some day automated machines may reach out to explore the distant star systems, to be followed by manned spaceships taking astronauts and later colonists through the zero of the starbow of relativistic flight.

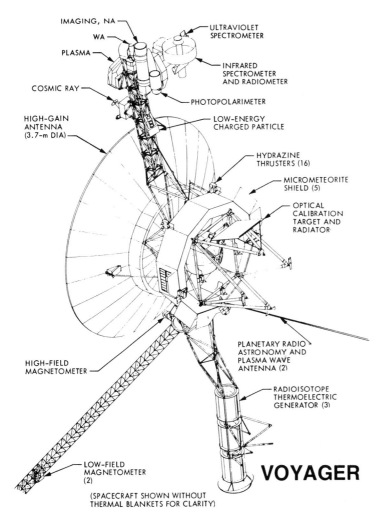

IMAGING, NA
WA
PLASMA
COSMIC RAY
HIGH-GAIN
ANTENNA
(3.7-m DIA)
ULTRAVIOLET
SPECTROMETER
INFRARED
SPECTROMETER
AND RADIOMETER
PHOTOPOLARIMETER
LOW-ENERGY
CHARGED PARTICLE
HYDRAZINE
THRUSTERS (16)
MICROMETEORITE
SHIELD (5)
OPTICAL
CALIBRATION
TARGET AND
RADIATOR
HIGH-FIELD
MAGNETOMETER
PLANETARY RADIO
ASTRONOMY AND
PLASMA WAVE
ANTENNA (2)
RADIOISOTOPE
THERMOELECTRIC
GENERATOR (3)
LOW-FIELD
MAGNETOMETER
(2)
(SPACECRAFT SHOWN WITHOUT
THERMAL BLANKETS FOR CLARITY)
VOYAGER

FIGURE 1.18
The structure and the science payload of the Voyager spacecraft are identified on this
NASA/JPL drawing.

The outer Solar System represents the next new frontier of human exploration. The first thrusts into which were the explorations of Jupiter and Saturn,
first by Pioneer spacecraft and then by the two Voyagers. Many fascinating
discoveries made at these two giant planets whetted the appetite for the subsequent mission of these spacecraft to Uranus and Neptune and outward to seek
the heliopause before drifting on to the stars.

The two Voyager spacecraft are identical (figure 1.18). Each is dominated by
a 12-ft (3.7-m) diameter dish antenna and extended booms for radioisotope
thermoelectric generators (RTGs), magnetometers, and a directable scan platform for those instruments that must be pointed very accurately. Various sub-

systems and electronics are contained within a central structure known as a bus.

Each Voyager weighs approximately 1820 lb (825 kg) and contains about 5 million equivalent electronic parts. An onboard computer detects faults and responds to protect the spacecraft during its long journey through space. The large antenna dish focuses the radio energy into a narrow beam for long-distance communications with Earth. It transmits at two frequencies, 8.4 GHz in the X-band and 2.3 GHz in the S-band (at wavelengths of 3.6 cm and 13 cm respectively). The X-band was used to transmit science and some urgently needed engineering data at high rates. The S-band was used to transmit routine engineering data about the state of the spacecraft at a much lower rate.

As mentioned earlier, electrical power cannot practically be generated in the outer Solar System by a reasonable area of solar cells because the intensity of solar radiation decreases as the square of the distance from the Sun. Instead the Pioneer and Voyager spacecraft use the heat generated by the radioactive decay of plutonium dioxide. In the Voyagers, the RTGs as they are called, produce about 400 watts of electricity. This type of highly reliable power supply had been proved in the Pioneer spacecraft that continued to function beyond the orbit of Neptune. Over time, of course, the output gradually decreases, but each spacecraft was designed to always require less electrical power than that which the RTGs can supply.

Each Voyager spacecraft is stabilized on its three axes in two ways—by gyro control for short periods of time during encounters, and by celestial control with star and Sun sensors for long periods of cruise. Each control system reorients the spacecraft as required by thrust from small hydrazine jets generated by attitude-control thrusters.

Instruments on the scan platform are located away from the bus so that they can be pointed over a wider range of directions, and can look backward around the large antenna dish when necessary. The scan platform is moved by electric motors operating through gears. These actuators slew the platform to point the instruments in azimuth and elevation very accurately.

A tape recorder within the spacecraft records data on those occasions when the information cannot be transmitted immediately to Earth; for example, when the spacecraft is hidden from Earth by a planet or a satellite, or when data are being gathered too quickly for the data transmission system to transmit to Earth.

Voyager has three computer systems; a Computer Command Subsystem consisting of two identical computers, a Flight Data Subsystem which prepares data for transmission by radio to Earth, and an Attitude and Articulation Control Subsystem to orient the spacecraft and to point the science instruments on the scan platform.

Commands from Earth to the spacecraft are sent by radio. At the spacecraft these commands are decoded and routed to the spacecrafts's Computer Command Subsystem where they are stored until required.

On April 6, 1978, not long after Voyager 2 was launched, its Computer Command Subsystem detected a failure and switched operations from the primary receiver to the spacecraft's backup receiver. But the telemetered engineering data showed that the backup receiver had a problem also. The receiver incorporates circuits that automatically track the frequency of the received signal as it varies owing to relative motions of spacecraft and Earth caused by what is known as the Doppler effect. As a result of the Doppler effect, the frequency of a received signal increases if the transmitter and receiver are approaching each other, and decreases if they are receding from each other. The frequency-locking circuit was not working on Voyager 2, so the spacecraft was commanded to return to the primary receiver. Soon afterward the primary receiver failed and the spacecraft automatically switched back to the crippled backup receiver. If this second receiver of Voyager 2 were to fail completely there would be no way of commanding the spacecraft for its encounters with the planets. To guard against such an eventuality a contingency backup mission load was sent to the spacecraft that would enable it to complete its mission in a somewhat limited way even if the receiver should fail and further commands could not be received by the spacecraft.

Because the receiver could only receive at one frequency, and changing the relative motions of Earth and spacecraft would cause changes in the frequency of the received signals at the spacecraft, it became necessary to vary the frequency transmitted from Earth to compensate for expected changes in frequency produced by the Doppler shifts. The problem was not completely solved, however, because temperature changes within the spacecraft could cause the receiver's frequency to change, and a temperature change of as little as 0.25 degrees C could shift the frequency by more than 96 Hertz (96 cycles per second) which would be outside the range that the receiver could respond to.

Voyager carries two classes of science instruments; target body sensors that are pointed in programmed directions, and field and particles sensors which detect the surrounding environment and do not have to be pointed.

The target body sensors consist of five subsystems; imaging science (ISS), infrared spectrometer and radiometer (IRIS), ultraviolet spectrometer (UVS), photopolarimeter (PPS), and radio science (RSS). The fields and particles sensors consist of six subsystems; planetary radio astronomy (PRA), plasma wave (PWS), magnetometer (MAG), plasma (PLS), low-energy charged particle (LECP), and cosmic ray (CRS).

The imaging science subsystem uses two camera systems—wide- and narrow-angle optical telescopes with slow-scan vidicons, each with eight selectable filters to record light at different wavelengths (colors). These cameras are used to obtain pictures of the planets and their atmospheric features, the ring systems, and the satellites. Imaging mosaics are used to map the surfaces of the satellites in as much detail as possible.

The infrared spectrometer and radiometer is also a camera whose telephoto lens and recording system are sensitive to near infrared and infrared radiation—that is, to heat energy or radiation at and beyond the red end of the visible spectrum; from 3000 to 500,000 angstroms. The IRIS measures the heat from planets and satellites and can be used to determine if certain elements and molecules are present in the atmosphere of the giant planets or on the surfaces of their satellites. The ultraviolet spectrometer records radiation beyond the violet end of the visible spectrum (from 500 to 1700 angstroms), to detect absorptions or emissions by certain atoms or molecules. These spectral effects are used to determine atmospheric compositions and physical processes taking place in these atmospheres.

The photopolarimeter measures the polarization of light reflected from the planets and their rings and satellites or scattered by atmospheric or ring particles. The instrument is a powerful camera system for the study of the texture and composition of surfaces and ring particles, as well as the composition of atmospheres. In addition, its very high magnification makes it an ideal instrument to record the changes in light during occultation of stars by planets, rings, and satellites. Changes in starlight, for example, as it is obscured by these bodies, provide details of atmospheres, ring particles, ring edge geometry, and satellite diameters.

The radio science subsystem is an active one, as opposed to all the other sensors which have to rely upon energy or particles to come to them. Radio science uses radio transmissions from the spacecraft picked up by the tracking stations on Earth. The passage of radio waves of two frequencies through planetary or satellite atmospheres provides details of these atmospheres as they bend and slow the radio waves. As the radio waves pass through rings they are changed and the observed changes allow one to estimate the number, size, shape, and thickness of the rings, and the sizes of the particles in them. The change in frequency of the radio waves as a result of Doppler effects allows the path of the spacecraft near a planet or satellite to be calculated very accurately. This, in turn, allows the mass of the planet or satellite to be determined.

Of the fields and particles experiments carried by Voyager into the outer Solar System, the planetary radio astronomy (PRA) subsystem measures radio waves from the Sun and planets, and the plasma wave subsystem (PWS) detects radio waves much like the PRA, but at much lower frequencies. The other four sensors make measurements of particles or fields depending upon the solar wind and the planetary magnetospheres.

The PRA is a sensitive receiver attached to two 33-ft (10-m) long antennas which detects radio signals produced by the Sun, the planets, the magnetospheres of the planets, and any lightning in the atmosphere of a planet. The PWS shares the same antennas as the PRA and samples the plasmas in and around planetary atmospheres by detecting radio waves generated in them. It also can detect radio waves generated by lightning flashes in planetary atmospheres, and the impact of very small dust particles on the spacecraft. The magnetometer measures the strengths of magnetic fields and their direction. The plasma wave (PW) subsystem detects electrically charged particles, while low-energy charged particles and cosmic ray subsystems detect energetic particles over various energy levels.

The names and affiliations of the principal investigators for these experiments for the encounter with Uranus are given in table 1.1. For the encounters with Jupiter and Saturn there were other principal investigators or team leaders for some of these experiments.

Associated with science experiments of Voyager there were on the average about 130 investigators including principal investigators and team leaders. Some of the teams changed for encounters with the different planets of the Voyager mission. These scientists were associated with universities, observatories, and aerospace companies in the United States, Canada, England, France, and Germany. Many students and research assistants also work for these investigators. A science investigation support team of 30 people helps to resolve conflicting requirements of the spacecraft for the different science experiments and their objectives. Another team of about 30 people attends to the sequencing of spacecraft science operations to keep experiments within the capabilities of the spacecraft.

An army of ground support people is essential for the success of the missions to the outer Solar System and to enable the scientists to make their investigations of the outer giants and their environments. A team of 10 navigators determines the location of the spacecraft throughout the mission and decides on the maneuvers needed to reach specific targets. They are assisted by a team of 10 orbit specialists whose job is to predict the positions of the planets and satellites so that the spacecraft can be navigated to rendezvous with them at preselected distances and times.

A spacecraft team consists of some 60 engineers who watch over the spacecraft and maintain it in a healthy operational condition. They examine engineering telemetry, analyze the returned data, develop alternative procedures to work around obstacles, and plan how the spacecraft should be operated to attain the mission objectives. They work closely with the sequencing team.

Another 40 engineers constitute a flight operations team. Their job is to operate the spacecraft on its long journey and to schedule the necessary ground support to do so. Major ground support came from Deep Space Network stations located around the globe and supplemented by Australia's Parkes Radio Observatory for the flyby of Uranus. The three Deep Space Network

TABLE 1.1
Voyager Science and Principal Investigators

Experiment & Acronym	Principal Investigator	Affiliation
Cosmic Ray (CRS)	Edward C. Stone	Caltech
Infrared Radiometry & Spectrometry (IRIS)	Rudolf A. Hanel	GSFC
Imaging Science (ISS)	Bradford A. Smith	U. of Arizona
Low-Energy Charged Particles (LECP)	S.M. Krimigis	Johns Hopkins
Magnetic Fields (MAG)	Norman F. Ness	GSFC
Plasma Science (PLS)	Herbert S. Bridge	MIT
Photopolarimetry (PPS)	Arthur L. Lane	JPL
Planetary Radio Astronomy (PRA)	James W. Warwick	Radiophysics
Plasma Wave (PWS)	Frederick L. Scarf	TRW
Radio Science (RSS)	G. Len Tyler	Stanford U.
Ultraviolet Spectroscopy (UVS)	A. Lyle Broadfoot	U. of Arizona

Caltech, California Institute of Technology
GSFC, Goddard Space Flight Center, NASA
MIT, Massachusetts Institute of Technology
JPL, Jet Propulsion Laboratory
Radiophysics, Radiophysics, Inc.
TRW, TRW Defense and Space Systems

stations are at Goldstone, California, Madrid, and Tidbinbilla, Australia. Each station has a 210-foot (64-m) diameter antenna for a radio transmit/receive station, and several smaller antennas. More than 500 people operate or support these stations, which have to be spread around the globe so that communications can be maintained throughout the day. The data received by the Deep Space Network stations are sent via the NASA Communications Network to the Jet Propulsion Laboratory at Pasadena, California, for processing, analysis, and delivery to users.

The encounter with Uranus required maximum collecting power of the signals on Earth. It was planned for when the spacecraft was in view of the Australian stations. Uranus as seen from Earth at the time of the encounter was in the constellation Scorpius. The maximum time the constellation, and therefore Uranus, could be above the local horizon was in the Southern Hemisphere. Picking the Australian sites for the encounter allowed an additional antenna area, and improved communications, because several antennas acted as one big antenna system. The Deep Space Network antennas at Tidbinbilla could work with the 210-ft (64-m) Australian radio astronomy antenna at Parkes to gather the extremely faint signals from the spacecraft.

Three antennas of the DSN station combined their received signals with that received by the Parkes' antenna through a 200-mile (320-km) microwave link. Without this combination, data could have been transmitted at only 10,800 bits per second, one quarter the rate at Saturn and one sixteenth the rate at Jupiter. The simultaneous tracking of Voyager by the four Australian antennas maintained a data rate of 21,600 bits per second for the relatively brief period of the encounter during which an enormous quantity of information had to be gathered at the spacecraft and transmitted to Earth.

Almost two years after the two Voyager spacecraft left Earth, they arrived at Jupiter; Voyager 1 made its closest approach on March 5, 1979, and Voyager 2 on July 9, 1979. The spacecraft confirmed that Jupiter is predominantly a gas giant consisting of hydrogen and helium approximating the abundances measured in the Sun. Helium was found to make up about 11% of the volume and 20% of the mass of the outer atmosphere of Jupiter.

The alternating patterns of light and dark zones and belts of Jupiter were imaged in great detail (figure 1.19) and their motions charted accurately to confirm wind systems and wind velocities. The highest wind speed recorded was 335 mph (540 kph) in an equatorial jet stream. Above the middle latitudes in both hemispheres, jet streams were seen moving in opposite directions adjacent to one another.

The Great Red Spot, which has been observed on Jupiter for several centuries from Earth, was revealed as an enormous anticyclone rotating in a period of about 6 days and carrying its clouds high above surrounding clouds. Other smaller white oval storm features also proved to be anticyclonic high pressure regions of rising atmosphere.

Lightning was detected planet wide at the tops of the clouds, and extensive aurorae were recorded at higher altitudes. Measurements of the heat radiated from Jupiter showed that the planet is still generating internal heat, by which the polar regions of its outer atmosphere are maintained at the same temperature as equatorial regions.

The Voyagers also discovered that Jupiter has a ring system. The rings, invisible from Earth, are extremely thin and consist only of very small particles, so they reflect hardly any light into the inner Solar System. They were discovered and observed by the Voyagers when the spacecraft were on the far side of the planet from the Sun. The rings extend from about two-and-a-half radii of Jupiter down to the top of the appreciable atmosphere, and the particles of which they are composed probably originate from small inner satellites.

Jupiter's intense magnetic field and radiation belts, first confirmed by Pioneer 10, were investigated again. The magnetosphere of Jupiter is ten times the size of the Sun and is the largest planetary magnetosphere in the Solar System. The extent of the magnetosphere is affected strongly by the intensity of the solar wind; when the solar wind is strong, the magnetosphere is compressed toward the planet. The major satellites of Jupiter orbit within the magnetos-

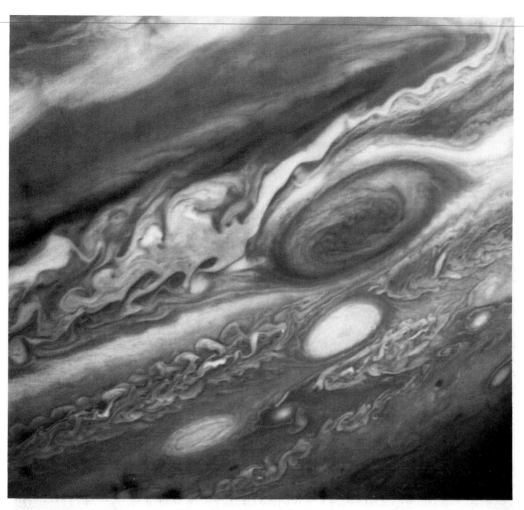

FIGURE 1.19
The alternating patterns of light and dark zones were imaged on Jupiter in great detail, and revealed the intricate structure of the Great Red Spot, which had been seen from Earth for centuries. (Photo. NASA/JPL)

phere so that interactions between satellites and magnetosphere are important. An electric current exceeding 5 million amperes flows along a magnetic flux tube connecting the innermost large satellite, Io, to Jupiter. Materials from the surface of Io form a hot torus of sulfur and oxygen plasma from which intense ultraviolet radiation is emitted and ions are accelerated to one-tenth the speed of light.

The four large Galilean satellites of Jupiter (figure 1.20) provided many surprises. Io displayed a number of active volcanoes, about nine of which were erupting during the flybys of the Voyagers. Some of the volcanic plumes spouted as high as 190 miles (300 km) into space. Material ejected from the volca-

FIGURE 1.20
The four large satellites of Jupiter provided many surprises including the volcanos on Io, the smooth surface of Europa, the tectonics on Ganymede, and the ancient cratered terrain on Callisto. (Photos. NASA/JPL)

FIGURE 1.21
Saturn was revealed in its awesome splendor. (Photo. NASA/JPL)

noes forms a plasma torus about Jupiter. Charged particles from this torus fall into the planet's outer atmosphere, and produce the brilliant aurorae observed there.

The satellite Europa, by stark contrast to Io, has a surface revealed by the Voyager images as smooth, uncratered, and fractured with many cracks. Speculation is that Europa is a water planet with a solid surface of ice, several miles thick. Ganymede, the next large satellite out from Jupiter, has a surface entirely different from either Io or Europa. It shows tectonic movement at crustal plates which have given rise to a variety of surface features. Many old craters are evident. Today they are degraded, and almost ghostlike in quality.

The outermost large satellite, Callisto, is different again. It has a heavily cratered terrain much like the ancient cratered highlands of Earth's Moon, with little or no evidence of any further geological activity since the craters were formed, probably several billion years ago.

Many new small satellites were also discovered orbiting Jupiter, bringing the total of known satellites to at least 16.

At Saturn (figure 1.21) further important discoveries were made about the planets of the outer Solar System. Voyager 1 made its closest approach to Saturn on November 12, 1980, Voyager 2 in May 1981—nearly two years after the encounter with Jupiter.

Like Jupiter, Saturn consists mainly of hydrogen and helium, with small amounts of methane, ammonia, phosphene, and a variety of hydrocarbons. The outer atmosphere of Saturn was found to have less helium than that of Jupiter (7% compared with 11%), possibly because the helium is sinking toward the core of the planet and in the process it is generating heat through gravitational separation. This would account for Saturn's emitting 80% more heat than it receives from the Sun.

Because Saturn is colder than Jupiter, a high-altitude haze screens cloud details. As a consequence, belts and zones are subdued in color and contrast, but the overall structure is similar to that of Jupiter. At the equator a high speed jetstream travels at 1060 mph (1710 kph), and as at Jupiter there are alternating jetstreams starting at mid-latitudes. A few anticyclonic storms were seen, but none close to the equator. Lightning flashes were observed close to the equator, and there were aurorae over a wide range of latitudes.

Saturn's extensive magnetosphere is also affected by the strength of the solar wind. The magnetic field of the planet is aligned within 1 degree of the axis of rotation. Two toruses are within the magnetosphere. A torus of hydrogen and oxygen ions probably originates from the satellites Dione and Tethys, and a torus of uncharged (neutral) hydrogen atoms probably originates from the atmosphere of Titan.

Perhaps the greatest surprise about Saturn was the complexity of the ring system (figure 1.22). The rings, seen as continuous from Earth, were found to be a complex system of many thousands of thin ringlets with apparent gaps between them. Moreover, the dark gaps seen from Earth were not voids, but contain faint ringlets. Also the apparent gaps between ringlets were later shown not to be completely free of material. New rings were discovered and the F ring, first seen on images returned by Pioneer 11, was confirmed, as was the presence of two small satellites shepherding the ring particles of the F ring and giving it a braided appearance at times. Elliptical and discontinuous ringlets were also identified. Clouds of particles extend above and below the ring plane to form spoke-like markings corotating with the B ring. These markings seem to develop within the shadow of Saturn and to dissipate as the ring particles move through the sunlit part of their orbits, not because the particles are in sunshine but because along a spoke they move at differing speeds (according to the law of Kepler governing the period of orbiting bodies at different distances from the primary). The spoke shape cannot be maintained.

The satellites of Saturn (figure 1.23) proved to be a complex group of small worlds. Some have enormous impact craters that must have come close to shattering the small worlds. Part of the surface of Enceladus is free of impact

FIGURE 1.22
One of the greatest surprises was the intricate nature of the ring system. (Photo. NASA/JPL)

craters, which suggests that there has been recent geologic molding of the surface. Iapetus has light and dark hemispheres thought to be caused by methane flowing over part of the satellite and being darkened in color by radiation effects.

Titan, the largest of the Saturnian satellites and a rival in size to Jupiter's biggest satellite, Ganymede, was observed at close quarters [2485 miles (4000 km)] by one of the Voyagers. In many ways it resembled a primitive and very cold Earth. Its dense atmosphere was probed to reveal several haze layers which prevented any imaging of surface features. A speculation is that Titan has oceans of liquid ethane, methane, and nitrogen.

New small satellites were also observed, bringing the total number in the Saturnian system to at least 17. The presence of very small satellites in the ring system is indicated by scalloping on the edges of some rings. This could be produced by perturbations from small satellites, too small to be identified on the images returned from Voyager.

The wealth of these discoveries and observations made it more important than ever that a spacecraft should reach Uranus and Neptune. Only one spacecraft, Voyager 2, could do so. In making a close approach to Titan, Voyager 1

FIGURE 1.23
The satellites of Saturn proved to be an extremely complex group; including Mimas almost shattered by an impact, Iapetus with dark and light hemispheres, lunar-like surfaces of some satellites, and internally molded surfaces of others. (Photos. NASA/JPL)

had been flung high above the plane of the ecliptic, beyond its capability to navigate to the outermost planets. Everything now depended on Voyager 2. There were no other plans for U.S. spacecraft to explore the outermost planets before the next century. The planets had moved out of a suitable alignment for using the gravity assist of Jupiter and Saturn, and direct voyages to Uranus and Neptune would take decades with available technology.

But Voyager 2 was an ailing spacecraft, with two serious problems. Only one command receiver was available, and it was suffering from a lack of frequency-locking capability, a fault that had occurred soon after leaving Earth. Additionally, the slewing mechanism for the scan platform had seized up soon after the closest approach to Saturn. The platform had subsequently been freed, but there was a haunting fear that it might seize up again at the critical period of encounter with Uranus or Neptune. So the ground support crews had to find ways to work around these two problems if successful encounters with Uranus and Neptune were to be achieved with any degree of certainty.

Voyager was also a sporadic target for budget cutting proposals. For example, when the spacecraft was halfway to Uranus in 1981, the White House Office of Management and Budget asked NASA to trim the agency's budget by $367 million for that year (Fiscal 1982), and by $1 billion in subsequent years. Cutting off Voyager was discussed as one way to contribute to these reductions. Fortunately this was not done.

As discussed earlier, the Grand Tour had been abandoned because of negativism in Congress about the value of scientific exploration of the outer Solar System deriving from apparent lack of long-term vision and long-range stimulating goals for the nation. But Voyager had the potential of reaching the outermost planets. This capability was downplayed early in the mission from, it seems, a political standpoint. NASA wished not to offend those in power who had been so opposed to the Grand Tour concept.

However, the visionaries at the Jet Propulsion Laboratory were working hard evaluating various orbits that the spacecraft might use and targeting options for planets, rings, and satellites. The enormously powerful simulation capabilities of modern digital computers were brought into play to model the many alternatives and to find out which were best. Scientists were consulted as to the most attractive scenarios for exploring the systems of Uranus and Neptune, and the navigators and space systems engineers worked out ways to achieve the important and challenging scientific objectives.

Voyager 2, as we have seen, could be navigated to fly through the system of Uranus and then on to Neptune. It was a great technical challenge, but it had the potential of obtaining priceless scientific information that might otherwise be denied us for a century. There was some argument that the space telescope would allow us to gather information about the outer planets without having to fly by them with spacecraft. This argument had some merits, but on the other hand it relied upon the space telescope being safely established in orbit in an operable condition. By contrast, Voyager 2 was already in space and well on its way to Uranus, and was certainly operable.

As it turned out, the Space Telescope could not be launched on schedule because the nation had unwisely committed itself to one mode of launching payloads into space—the space shuttle, basically a test machine, which was pushed as an operable transportation system. It inevitably failed under pressure to follow an unreasonable turnaround and launch schedule.

Outer planet exploration, as well as all other American space missions, experienced a devastating blow because of this unrealistic national policy; the in-depth exploration of the Jovian system by the highly sophisticated Galileo space system as well as the telescope in space program was pushed back into the 1990s. The outcome was that everything ultimately depended on keeping the

Voyager 2 operating in space for a journey beyond Uranus to Neptune, and keeping the Voyager 1 and the two Pioneer spacecraft operable to return data about the interplanetary environment as they pushed on toward the heliopause.

Voyager 2 could be commanded to follow a trajectory past Saturn that would allow the spacecraft to gather supplementary or confirmatory scientific data and then proceed to Uranus and Neptune. The original objectives of the Voyager missions could then be expanded to include gathering scientific data about Uranus and Neptune by flybys of those very distant planets. The Grand Tour could essentially be reinstated, at least, in part. Only Pluto would remain unexplored. Voyager 1, its mission accomplished, headed in the general direction of the outermost and most mysterious planet, but would be millions of miles above Pluto's orbit with no way of making a flyby.

In February 1984 a workshop was held at the Jet Propulsion Laboratory, California to discuss the scientific aspects of the exploration of Uranus and Neptune, and four major study objectives were identified—investigating atmospheres, rings, satellites, and magnetospheres. Once the scientists had defined the objectives in detail—for example, the composition and structure of the atmosphere, the global energy budget, the circulation patterns, and so on—observations could be defined that would be needed to collect the necessary data. Then a mission plan was evolved to make sure that the spacecraft and its instruments were configured and commanded so as to make the observations. This involved selecting the trajectory and the timing of the encounters with the planet and its satellites, arranging for occultations by timing and positioning the trajectory, and resolving conflicts between the requirements for different scientific observations. Additionally, the operations at Uranus were constrained by the need to continue to Neptune. The encounter with Neptune had to be arranged so that the spacecraft could fly close to the planet and to the large satellite Triton, with Earth occultations by both bodies and opportunities to resolve questions about Neptune's ring system. Finally the plan had to be converted into a set of computer commands that could be sent to the spacecraft and loaded into its memory.

But Voyager 2 had problems, as discussed earlier; a receiver that could not lock on incoming signals from Earth, and a scan platform actuator that had seized up. Fortunately there was plenty of time to develop ways to circumvent these problems. After the encounter with Saturn nearly 4½ years would elapse before Voyager 2 reached Uranus. An extensive program of testing of the spacecraft in flight and of similar equipment on the ground brought more understanding of the problem with the sticking actuator. It appeared to be a problem of lubricating the gears. However, if the actuator was commanded to move slowly, it worked fine. It seized up only when attempts were made to slew the platform rapidly, as was done at Jupiter and Saturn. So for the Uranus and Neptune encounters, sequences were planned to use slow slew rates only. A medium slew rate would, however, be permitted when absolutely necessary to obtain data that could otherwise not be gathered. If the actuator should seize up completely, alternative plans had been developed to move the entire spacecraft by means of its thrusters so as to point the instruments in the required directions during the encounters with Uranus and Neptune. This alternative plan was incorporated into commands stored within the spacecraft. The intention was to test the scan platform about 100 hours before encounter with Uranus, and if the actuator failed, to switch to the contingency plan. The spacecraft was also given a set of commands for scientific observations should the receiver fail and sever the command link to Earth.

There were, in fact, five failure modes studied. The remaining receiver might fail, thus ending the ability of the spacecraft to receive commands from Earth. The workaround for such as failure was the stored set of commands. The scan platform might seize up and be unmovable in azimuth. If so, roll turns developed by the thrusters would be substituted for the azimuth slews. The scan platform might become immovable in elevation. This was considered unlikely, so no backup mode was developed. One of the two computer command subsystem processors on the spacecraft might fail and a backup command load had been prepared for such an occurrence. Finally, one of the two flight data systems might fail. Again, a backup command load had been developed. Otherwise everything on the spacecraft and the ground system was operating as expected and the encounter with Uranus seemed assured.

The only major technical problem was one of physical limitation; that of obtaining the data at the low light levels in the Uranian system, and transmitting it back to Earth over the enormous distance. Uranus, as noted, is a system of darkness. Additionally, there were uncertainties of position caused by the great distance to Uranus and the relatively short period over which the motions of the satellites, particularly Miranda, had been observed since their discovery.

The exposures for observations that rely on sunlight—e.g. images of the satellite—had to be increased considerably over those used at Saturn. Long exposures usually require that the camera and the subject remain relatively motionless during the exposure, but the spacecraft carrying the camera would be veritably hurtling through the Uranian system at 45,000 mph (72,500 kph). Smeared images would result, with consequent loss of detail unless the relative motion could be compensated for. Two solutions were developed for the smear problem. First, the spacecraft was held steadier by reducing torques from other subsystems. For example, when the tape recorder was started or stopped it induced a small compensatory rotation of the spacecraft. This rotation was prevented by firing the hydrazine thrusters in the appropriate directions at the time the tape recorder started or stopped. The thrusters, too, had been commanded to fire at only half of the thrust used during the Saturn and Jupiter encounters, thereby permitting much smoother control of the spacecraft's attitude at Uranus. The second solution was called image motion compensation and was achieved by maneuvering the whole spacecraft to compensate for the relative motion between camera and subject as the spacecraft sped close by Miranda. This maneuver would, however, move the antenna so that it would point away from Earth. Communications would be interrupted until the spacecraft could be maneuvered again after the encounter with Miranda. The whole process had to take place with minimum interruption of the radio signal link to Earth so that Doppler data on the signal from Voyager could be measured to determine how Miranda changed the path of the spacecraft and thereby to determine the mass of Miranda.

Another technical problem of physical limitation was the capacity of the telemetry system to send data to Earth. The inverse square law applies to communication rates, with the consequence that the number of bits of information that could be transmitted from the spacecraft at Uranus to Earth each second was considerably reduced. This posed problems for the imaging experiments since it would reduce the number of complete images that could be gathered and relayed to Earth. To overcome this problem the imaging data were compressed on the spacecraft so that only differences between adjacent picture elements (pixels) of an image were transmitted, a technique which had been applied many years earlier during Mariner 10's close encounter with the planet Mercury and had proved very successful. About twice as many images could be sent from Uranus to Earth in this mode than would have been possible if the technique of transmitting each pixel had been used, as it had been at Jupiter and Saturn. Moreover, as already mentioned, the ground collection capabilities of the Deep Space Network had been improved by arraying several antennas to collect and combine the very faint signals received at Earth from the spacecraft.

At Uranus' distance—2 billion miles from Earth—it takes 2 hours and 45 minutes for light and radio waves to travel to Earth. It takes 90 minutes to transmit a sequence load of commands to the spacecraft. So before the first element of a command load sequence reached the spacecraft, the last element in the command sequence would have been transmitted from Earth. The command sequence would thus consist of a long string of radio pulses stretched out in space and moving at the speed of light toward Uranus. It took about 5½ hours from the time the last part of the sequence was transmitted from Earth for a signal to get back to Earth stating that the command sequence had been received by the spacecraft. So every activity of the spacecraft had to be planned carefully well in advance of when it would be required.

Nine command sequences had to be stored in the spacecraft for the encounter with Uranus. Had the spacecraft failed in any of the modes discussed earlier, three contingency sequences were ready for transmission to the spacecraft to allow for a failure of the slew activator, a failure of a computer command subsystem processor, or a failure of a flight data system computer. But this could be done only if there was sufficient time to become aware of the failure and get the command sequence to the spacecraft before the revised command sequence had to operate.

Before the actual encounter with Uranus, a number of functions of the spacecraft had to be tested in a pre-encounter test sequence. Additionally, the Deep Space Network and the Ground Data System had to be tested for its configuration and operational readiness. On the spacecraft, calibration tests were made of the science instruments. During this testing period a sequence of maneuvers was executed, the radio science occultation equipment was tested, the power available in the spacecraft was evaluated, the infrared spectrometer and radiometer and the imaging science subsystem were calibrated by viewing a target plate carried by the spacecraft, and wide-angle images of star fields were analyzed to determine how accurately the spacecraft could be pointed in a desired direction.

This test sequence was completed by early November 1985 with the spacecraft on a bull's-eye approach to the Uranian system. The encounter consisted of four phases: an Observatory Phase from November 4, 1985 to January 10, 1986; a Far Encounter Phase from January 10 to January 22; a Near Encounter Phase from January 22 to January 26; and a Post Encounter Phase from January 26 to February 28.

During the Observatory Phase, Voyager obtained the first images of Uranus with better resolution than photographs which can be obtained from Earth. The plan was to obtain sequences of images from which, as with Jupiter and Saturn, long-term atmospheric motions could be monitored to show movements of markings around the disk of the planet and establish periods of rotation and chart wind systems. Atmospheric features would be tracked and a search would be made for aurorae. Particles and fields of interplanetary space would also be monitored, and emission of ultraviolet from hydrogen in space between the satellites and the planet would be searched for. Between November 27 and December 23, Uranus would be at superior conjunction on the far side of the Sun from Earth. As a consequence, radio waves from the Voyager would pass close to the Sun, so that the effects of the Sun and its atmosphere on their transmission could be studied.

During the Far Encounter Phase imaging mosaics would be gathered of the planet, its rings, and the satellites. Observations by the infrared spectrometer and radiometer would determine the global heat balance on the planet, and optical navigation images would be obtained for precise positioning and navigation to establish more accurate ephemerides, and particularly to refine the targeting for the close encounter with Miranda. Final corrections would be made to the target plane aiming of Voyager. Toward the end of the phase critical activities would take place, including a final torque margin test. This test of the scan platform at 4.5 days before closest approach would ascertain if the platform were functioning correctly and could be used during the encounter rather than having to orient the spacecraft itself. If the scan platform did not pass the test, the contingency commands would be transmitted to the spacecraft. A late ephemeris update was made to the pointing routines for the scan platform needed during the encounter, and late updates from the final optical navigation results were provided for the maneuver needed to compensate for relative motion between satellite and cameras when the spacecraft would later hurtle past Miranda.

The Near Encounter Phase was one of much activity during which the highest resolution images of the planet, rings, and satellites were obtained. The magnetosphere structure was investigated, masses of Uranus and Miranda were determined, radio emissions from Uranus and the rings were checked, and various occultations were used to obtain high resolution details of the atmosphere of Uranus and details of the ring system. Earth occultation observations used the radio science subsystem, star occultation observations used the photopolarimeter subsystem. Sun occultations were observed with the ultraviolet spectrometer, and atmospheric near-polar occultations were observed with the ultraviolet subsystem and imaging of a relatively bright star, gamma Persei.

During this phase a number of maneuvering and roll sequences were commanded to meet the target sequencing and image motion compensation requirements.

The Post Encounter Phase experienced a gradual reduction of activities. It began with high phase-angle measurements of the planet and its ring system, and a search for additional backlighted rings. Phase angle is the angle subtended between the direction of incident light on the subject being observed and the direction to the observing instrument. A low phase angle is when the viewpoint is almost the same as the direction of the lighting; a high phase angle is when the viewpoint is almost toward the direction of the lighting. Faint rings invisible on approach were expected to be revealed at high phase angles by their forward scattering of sunlight. The magnetotail of Uranus was explored, and maneuvers took place to direct the spacecraft for its later rendezvous with Neptune. Finally the imaging and infrared subsystems were calibrated again, and the spacecraft was rolled to use Achernar (the end of the river), the brightest star in the constellation Eridanus, as a reference star.

Voyager 2 would then be on its interplanetary cruise to the final planetary target, Neptune—a journey of another 1.7 billion miles beyond the orbit of Uranus that would last 3½ years. But Voyager 2 had added an enormous amount of new information about the Uranian system that would allow scientists to plan for the Neptune encounter and refine their views of what must be sought for at the outermost of the large planets. The discoveries at Uranus were in general fascinating, and many were quite unexpected.

2
GEORGIUM SIDUS

For almost two millennia our views of the universe were circumscribed by a series of transparent spheres surrounding the Earth and carrying all the celestial bodies. This concept accounted for the apparent paths of the planets across the night sky, believed to be the true paths of these bodies.

The universe was thus centered around the Earth, which was surrounded by the concentric spheres. The innermost sphere carried the Moon, and then, in order of increasing distance, the other spheres carried Mercury, Venus, the Sun, Mars, Jupiter, Saturn, the fixed stars, and finally, God and His Saints. Moreover, this universe was accepted as being complete and unchanging.

The meticulous observations of celestial motions by Tycho Brahe in the sixteenth century, followed by the interpretations of Johannes Kepler and Nicolaus Copernicus that the planets—including Earth—orbit the Sun, disturbed the ancient hierarchy of the celestial spheres. The invention of the telescope and subsequent observations of Galileo Galilei confirmed the Sun-centered Solar System. But although ancient Burmese writings mentioned another invisible planet named Rahu, completeness was still assumed. A disturbance of the mystic seven heavenly bodies—Moon, Sun, Mercury, Venus, Mars, Jupiter, and Saturn—was deemed to be heretical to Europeans, and books by emerging astronomers and natural philosophers that stated otherwise were proscribed and their writers were excommunicated and in some instances put to death.

However, almost a century before the last of the banned books was finally removed from the *Index Librorum Expurgandorum,* a son was born to a poor musician in Hanover, Germany, on November 15, 1738, who was destined to change our entire view of the heavens. William Herschel widened the vistas of the universe as would Charles Darwin of the living systems within that universe.

Uranus was discovered in the eighteenth century by this amateur astronomer and professional musician. The discovery resulted from Herschel's attention to detail and his painstaking effort to catalog the heavens. He added to the Solar System the first planet to be discovered in recorded history.

William Herschel became a musician like his father and he joined the military band of the Hanoverian Army. The military life did not suit him. After the defeat of the Hanoverian troops by the French at Hastenbeck, Herschel fled to England to start a new life there.

Today if you take the Motorway M4 from London toward Bristol and the West Country, you soon pass Slough, where Herschel died in 1822, and in a little more than 90 minutes driving at the legal speed limit of 70 mph you come to exit 18 and the Bath road. Descending a steep hill you meet the old London road from Swindon, turn right and enter the town of Bath. Here it was where Herschel made his discovery of Uranus on March 13, 1781. He is commemorated by a museum established at the house in King Street from which he first saw the planet.

By the mid-1700s Bath had suddenly blossomed from a backward town of a few thousand country people that had virtually stagnated for centuries following the Roman abandonment of their *Aquae Solis,* with its magnificent baths and temple of Minerva. As part of the stimulus to provincial towns encouraged by George III's reign, Bath quickly evolved into the equivalent of a Las Vegas to eighteenth century England. Aristocrats and the *nouveau riche* of Britain's mercantile marine trading and the industrial revolution gravitated to the rapidly expanding town on the River Avon to gamble and to be entertained. Fashionable Georgian houses were built in crescent-shaped terraces to accommodate the influx of people. Bath became a natural attraction for a struggling musician who had tried for a living in Midlands and Northern cities (London presented too much competition for musicians) without much success. Herschel moved to Bath in December 1766 and there built up a flourishing business, performing himself and teaching others. He developed an income of about £400 a year, and brought brothers and a sister from Hanover to live with him.

With his finances in order, he was able to follow his astronomical hobby. He rented property in King Street, not far from the heart of Bath, which provided easy access to his musical activities at the Octagon Chapel, a private center of religious worship where the fashionable elite could avoid having to mix with the masses who worshipped at the Abbey. The King Street property also had facilities for his astronomical work. Subsequently he moved several times in and out of King Street during the years he spent at Bath.

Herschel's astronomical aims were directed toward the important question then puzzling astronomers; how distant are the stars? Herschel had been an avid reader of astronomical and mathematical books for years, but he was not content merely to read. He wanted to see the astronomical bodies for himself. The crude refracting telescopes he rented for this purpose were disappointing to him to the point of frustration. They were extremely long and difficult to manage for steady viewing. Their great length was needed to minimize chromatic aberration; colored flares around objects viewed through the telescope which were the result of glass refracting light of different colors by different amounts. The different colors could not all be brought to the same focus by the simple lenses available at the time, and therefore images of astronomical objects could not be magnified by an eyepiece without giving rise to the objectionable colored fringes. With a lens of very long focal length, light rays were not refracted as much as for a lens of short focal length, and colored fringes were reduced. So some of these telescopes were made with lengths of hundreds of feet—totally impractical for observation except on the most wind-free nights.

About 100 years earlier James Gregory had offered a solution to the problem. Use a mirror instead of a lens, because a mirror relects light of all colors in the same way. Herschel knew of the Gregorian reflector and of Isaac Newton's practical version of it, and so, rejecting the use of refractors, he accepted the mirror as providing the key to making practical bigger and better telescopes. He seems to have ignored the fact that John Dolland of London had discovered in 1757 how to make a compound lens of flint and crown glasses that could overcome chromatic aberration and considerably improve the refracting telescope. But at the time of Herschel's activities the new achromatic lenses could not be fabricated greater than 4 inches in diameter and this may be why Herschel did not use them. He sought after very large apertures for light gathering power. The aperture of a telescope is expressed in terms of the diameter of the primary mirror or lens which brings the image of a distant object to a focus where it is enlarged by an eyepiece. More aperture provides more light and allows higher magnification so that close objects can be separated.

Herschel rented a Gregorian reflector in September 1773 and very soon afterward bought the tools and other equipment to make his own reflecting telescope, since to have a professional make a mirror of the aperture he needed was beyond his financial means. While mirrors for reflectors are today made of glass, in Herschel's time they were of speculum metal—an alloy of tin and copper. In his mirror casting and finishing he was assisted by his brother Alexander and by his sister, Caroline. Almost all the rooms in the three-story house on King Street were turned into workshops. After two hundred failures and periods when he would keep his hands on a mirror for 16 hours of uninterrupted polishing, he completed a 5-inch-diameter mirror. By March 1774 he was recording observations of Saturn's rings and other astronomical objects. He observed a spot on Saturn from which he was first to determine the rotation period of 10 hours 16 minutes for that planet.

Herschel next confounded the contemporary expert optical workers and produced telescopes able to penetrate further into space than ever before. With these larger telescopes, he discovered two additional small satellites of the planet, Enceladus and Mimas. Except for reading a rudimentary book on optics, he developed his skill at telescope making by teaching himself through trial and error supported by an indomitable perseverance.

Came the summer of 1774 and Herschel moved to another house on the outskirts of Bath, with more expansive facilities and a flat roof that served admirably as an observing platform. The size of the reflectors continued to increase, because he needed larger apertures since he had turned his attention to the problem of stellar distances.

Theoretically one should be able to measure the distance of a star by simple trigonometry. Look at it from two ends of a baseline and measure the angular difference of the viewpoint and you can calculate the distance in terms of the baseline, which you can easily measure. This can be illustrated with a simple example. If you stand where you can see an object in the foreground against a distant background you can make the close object appear to move from side to side in front of the background by first closing one eye and then the other. The distance between your two eyes is the baseline, and the displacement of the foreground object provides the angular measurement. But suppose you do not have an object close enough to act as a foreground object. Then, the distance between your eyes is much too small for a baseline. With more distant objects, for example a distant copse of trees in which one tree has to be the foreground object, the measurement of angular displacement is not so easily obtained. One has to move an appreciable distance from side to side to have a baseline long enough to see the displacement of one distant object compared with a more distant one.

Astronomers had used this principle to measure the distances of nearby bodies such as the Moon, by observing the Moon at the same time from positions remote from each other on the Earth's surface. The stars are so far away, however, that the Earth itself is not big enough to include a long enough baseline on its surface. But the Earth travels around the Sun in one year, so that by making observations six months apart, an astronomer can use a baseline equal to the diameter of the Earth's orbit, namely about 186,000,000 miles (300,000,000 km). With this baseline a star 19,150 billion miles (3.26 light years) away from Earth will be displaced 2 arc seconds. This distance is termed a parsec. At the time of Herschel no one had determined the distances of the stars using the baseline of the Earth's orbit, although there had been attempts to do so. This failure indicated that the stars must be at enormous distances from Earth. Today we know that even the closest star is 4½ light years way.

Herschel was convinced that the lack of success could be attributed to lack of instrumental power in observing the stars. He thought that if he could observe stars with a very powerful telescope they would show disks that would help to determine their distance. But if they could not be observed as disks then he might be able to see the difference in position if he concentrated high telescopic power on stars that appeared very close together in the sky and whose angular separation could be measured within the same high-magnification telescopic field of view six months apart. So he started measuring the angular separation of all the close double stars he could find. At the same time he also pushed for bigger telescopes to see if he could resolve any star into a disk instead of the mere point of light that all the telescopes so far had shown. In addition he catalogued the stars to see if any had changed position over the years as a result of being closer than other stars.

So Herschel had two consuming projects: observe and map the heavens—inspecting all stars for any sign of a disk—and measure the angular distances between close doubles six months apart to check if there was any measurable change.

It was thus no accident that Herschel discovered Uranus. It was a direct and inevitable result of his painstaking cataloguing and astute observations of stars. In 1781, at 17 King Street—to which he had then returned—Herschel was busy examining and cataloguing stars in Gemini and Taurus for his second review of the heavens. He recorded an unusual star which he described as "one that appeared visibly larger than the rest." Herschel had an advantage over his contemporaries in that he had made telescopes capable of using eye-pieces of great magnifying power. He was able to identify the object as having a disk, whereas other astronomers who had earlier seen the object had recorded it as a star.

Herschel thought at first that the object was a comet. He described it to the Bath Philosophical Society; "in the quartile near ζ Tauri the lowest of the two [stars] is a curious either nebulous star or perhaps a comet." But he acknowledged that it was a strange comet at that. He communicated his discovery with the British Astronomer Royal, Nevil Maskelyne, and news of his observation spread rapidly to other astronomers, who began to track the strange object's motion relative to the stars. Subsequent observations showed that its path was almost circular, in contrast to the parabolic or highly elliptical paths of comets. It was not strange that Herschel did not recognize the object as a planet. There was no reason for him to have done so; everyone at that time regarded the known Solar System as complete. But as the newly discovered object continued to be tracked it became obvious that the boundaries of the Solar System were far beyond the orbit of Saturn.

Mathematicians of the time, who had been stimulated by the works of Kepler and Newton, found the movements of celestial bodies challenging exercises for their mathematical skills. They quickly tried to calculate the orbit of the new "comet," but found that it did not fit into the usual pattern. By about three months after Herschel's discovery, a Russian, Anders J. Lexell, at St. Petersburg (later renamed Leningrad), and a Frenchman, Pierre Simon Laplace, Professor of Mathematics at the Ecole Militaire, Paris, independently recognized that the mysterious object was indeed a planet. Its motion could be accounted for only by assuming that its orbit was almost circular at an undreamed of distance of twice as far as Saturn and about 19 times the distance of Earth from the Sun. Its period of revolution was about 84 years. The absolute number seven of the ancient planets was proved mythical.

Herschel was buried at Saint Lawrence churchyard at Upton on the outskirts of Slough where he had lived later in his life. On his tombstone were the words; "*Caelorum perrumpit claustra.*" He did, indeed, break through the barricades of restricted thought about the heavens, doubling the size of the Solar System and pushing men's minds out toward the distant stars by demonstrating their enormous distances.

During his lifetime Herschel referred to the new planet as Georgium Sidus, a name he bestowed upon it in honor of King George III of England who appointed Herschel King's Astronomer following his discovery of Uranus, and provided him with a yearly stipend of £200. Continental astronomers, who saw no reason for naming it after the king of another country, disagreed with Herschel's name for the new planet. The French astronomer Joseph J. Lalande, professor of astronomy at the Collège de France, proposed the new planet should be named Herschel, but other continental astronomers selected the name Uranus. In the mythology of ancient Greece, Uranus, the most ancient of the gods and their ruler, was the father of Saturn. This name was in keeping with the names of all the other planets and ultimately prevailed.

An interesting result of this naming process was that a German chemist, Martin H. Klaproth, who had just discovered the very heavy radioactive element uranium, and had named it klaprothium, changed the name of the element to uranium in honor of the discovery of the new planet. Ultimately uranium would lead to the development of nuclear power, a product of which was the plutonium used to produce electrical energy for the Voyager spacecraft that would explore Uranus.

Before Voyager, Uranus remained a very mysterious world about which little was known. The problem was that its great distance hampered observations from Earth. At 19.2 astronomical units from the Sun, a distance of 1606 million miles (2585 million km) at its closest approach to Earth, the disk of the planet subtends an angle of only 3.76 seconds of arc, compared with 45 seconds of arc for the disk of Jupiter and 25 for Mars at its closest approaches once every 15 years.

When viewed by a large telescope under good seeing conditions, when Earth's atmosphere is steady, Uranus appears as a fuzzy blue green and unmarked disk. Unlike Jupiter and Saturn the planet appears quite featureless,

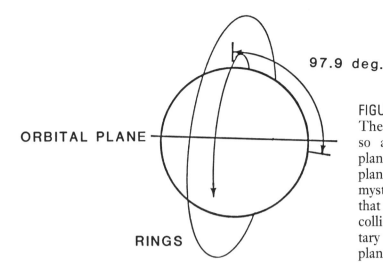

97.9 deg.

ORBITAL PLANE

RINGS

FIGURE 2.1
The axis of Uranus is tilted so as to lie almost in the plane of its orbit. Why the planet is so oriented is a mystery. One speculation is that there was a catastrophic collision with another planetary body early in the planet's history.

although astronomers have from time to time claimed to have seen a belted structure; a light equatorial belt with dark belts on either side. Other astronomers claimed to see dark and light patches but no zonal markings. The limb of the planet exhibits some darkening in comparison with the rest of the disk which implies that the planet does have an atmosphere.

In recent years the use of charge-coupled devices on large telescopes to electronically amplify the light from the planet revealed a dark polar area but no other details. Earth-based observations did, however, establish the fact that the planet differs markedly from the other planets of the Solar System in that its axis of rotation lies close to the plane of its orbit around the Sun; Uranus rotates on its side (figure 2.1). Its axis is tilted 98 degrees. The rotation rate was much in doubt. Various determinations using different methods derived rates between about 10 and 24 hours.

The diameter of Uranus is about four times that of Earth. In the 1950s the diameter of the planet was measured as 29,392 miles (47,300 km), but this was refined in the 1970s to 31,570 miles (50,800 km). By the time of Voyager's encounter it had been further refined by Earth-based measurements to 32,000 miles (51,500 km).

The density of the planet was known to be about 1.36 times that of water, which implies it is different from Jupiter and Saturn. Uranus was visualized as a planet consisting of an Earth-sized rocky core surrounded by a deep ocean of water topped by a 5000 mile (8000 km) deep atmosphere of hydrogen, helium, methane, and ammonia.

Five satellites were known, all difficult to see except in telescopes of large aperture. Two of these, Oberon and Titania, were discovered by Herschel in 1787 using a mirror of 18.7 inches aperture. Two more, Ariel and Umbriel, were discovered in 1851 by William Lassell, an English astronomer born in

Bolton, Lancashire, who also built his own speculum mirrors and observatories. He used a mirror of 24 inches aperture. The fifth satellite was not discovered until fairly recently (February 16, 1948) when Gerard P. Kuiper photographed Miranda with the 82-inch reflector at the McDonald Observatory in the United States. Herschel's son, John, suggested the names for the first four satellites, and Kuiper selected the name for the fifth. All names are characters from Shakespeare's plays.

The sizes of the satellites cannot be determined with great precision from Earth and depend very much on the albedo accepted for the star-like bodies; a darker surface would make a satellite appear fainter than would a bright surface and this in turn could be interpreted as due to a smaller diameter. Since water ice was detected on the surfaces of all the satellites, the calculated diameters after this discovery were smaller than beforehand, because water ice was assumed to produce a bright surface. The larger of the satellites, Titania and Oberon, have diameters less than half that of the Earth's Moon. More recently the surfaces of the satellites were shown to be darker despite the presence of water ice, so their diameters had to be revised upward.

Just before Voyager's encounter with Uranus the approximate sizes of the satellites were accepted as: Miranda 310 mi (500 km); Ariel, 825 mi (1330 km); Umbriel, 690 mi (1110 km); Titania, 995 mi (1600 km); and Oberon, 1010 mi (1625 km). Their respective densities were calculated as 3, 1.3, 1.4, 2.7, and 2.5, which was an unusual distribution compared with the Jovian and Saturnian systems, where densities increase inward toward the planet. The satellites were believed to have rocky cores with icy crusts of various thicknesses, thinnest for Titania and thickest for Ariel, with the composition of Miranda very uncertain. The less dense satellites would be expected to have deep mantles of icy materials, whereas Titania and Oberon would be mainly rocky with a relatively thin crust of icy materials. The surface temperature of all the satellites appeared to be about $-315°$ F. ($-193°$ C.).

Details of these satellites based on angular measurements, brightness, and assumed albedos before the Voyager encounter are summarized in Table 2.1.

The question was why the surfaces should be darker than the icy satellites of Saturn when water ice was expected there. One explanation as that the catastrophic event which turned Uranus onto its side might also have splashed dark rocky material into space, later to be accreted to the surfaces of the satellites. Only a relatively small amount of dark material is needed to darken the surface of an icy satellite. Another plausible explanation for the source of the dark material was that a radiation environment trapped particles in a magnetosphere of Uranus, and might have converted methane ices into complex organic polymers which are dark in color.

Recent work with radiometry and spectrophotometry from Earth revealed some of the basic physical properties of the Uranian satellites before the Voy-

TABLE 2.1
The Known Satellites of Uranus Before Voyager

Name	Orbit Radius mi	Orbit Radius km	Period (days)	Size Diameter mi	Size Diameter km
Miranda	80,000	130,000	1.41	310	500
Ariel	119,300	192,000	2.52	825	1330
Umbriel	165,900	267,000	4.14	690	1110
Titania	272,200	438,000	8.71	995	1600
Oberon	364,100	586,000	13.46	1010	1625

ager encounter. Radiometry measures the intensity of radiant energy by its heating effect and spectrophotometry compares the radiant power received at different regions of the spectrum. Using the radiometric technique to analyze the radiation received at Earth from the Uranian satellites allowed the albedos to be more accurately determined and hence to recalculate the diameters of the larger satellites. Spectrophotometry suggested that Miranda had the brightest surface of all the five satellites, thereby placing its diameter at about 300 mi (500 km), but with an uncertainty of at least plus or minus 100 miles. All the satellites appeared to have water ice surfaces mixed with darker material of unknown composition. The spectral signature of this dark material might be that of charcoal according to some observations, or possibly that of material similar to carbonaceous chondritic meteorites, a somewhat rare class of stony meteorites containing water and grains of hydrocarbon compounds including amino acids. But there was really no strong, definitive, spectral signature to identify the dark material. Close matches between laboratory spectra of mixtures of charcoal and water frost and those of the Uranian satellites have been obtained.

An important feature of the satellites of Uranus is that they are the most distant of the regular satellite systems; Neptune, as is discussed in a later chapter, does not appear to have a regular system of satellites. Uranus is thus more akin to Jupiter and Saturn than to Neptune as far as satellites are concerned.

The Uranian satellites also move in very regular orbits extremely close to the plane of the planet's equator. Unfortunately for observations of the satellites from Earth with the new instruments available during the decade before the Voyager encounter with the planet, the Uranian system was oriented with the poles of the planet, the satellite orbits, and the satellites facing toward Earth. Consequently it was difficult to make observations that would reveal albedo differences between satellite hemispheres. The aspect did, however, make it

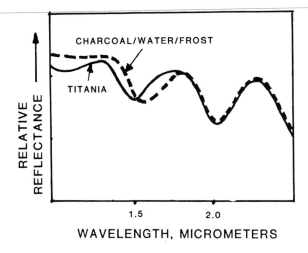

FIGURE 2.2
This graph shows a smoothed spectrum of Titania compared with one for water frost, showing the similarity. Methane ice and water ice both appear to be present on the surface of the Uranian satellites.

better for determining orbital parameters of the satellites and thereby checking interactions to determine the masses of the satellites.

To find out what the composition of the surface materials might be, scientists examined the spectrum of radiation reflected from the surface in the near infrared region—i.e. just beyond the red end of the visible spectrum. In some parts of this spectrum, the intensity of radiation was reduced by the surface materials absorbing it. These absorption features of the spectrum related to water ice or frost. The surface of the innermost satellite, Miranda, could not easily be checked in this way because of its proximity to Uranus and because its small size made its image so faint. Water ice spectra of Miranda were, however, obtained in 1984, and it became clear that water ice is present on the surfaces of all the Uranian satellites.

Evidence appeared to be mounting that whereas the surfaces of the larger satellites were heavily contaminated, possibly by infall of meteoric material, that of Miranda had relatively pure water ice, similar to some of the icy satellites of Saturn, for example. A smoothed spectrum for Titania relative to one for water frost is shown in figure 2.2. There were suggestions of the presence of ammonia, methane, and carbon monoxide in the spectra of the Uranian satellites which led to the speculation that the dark material might be methane converted to carbon and a dark polymer by radiation.

The density of a satellite is an important physical property needed to understand the interior of the satellite and how it may have formed. Until very recently this density could not be determined for the Uranian satellites because their radii and their masses were very uncertain. However, recently radiometric observations comparing the radiation at 20 micrometers with that in the visual region were used to determine the radius and albedos of the satellites, and estimate densities (see table 2.2). By assuming densities and albedos, masses of

TABLE 2.2
Density of Uranian Satellites From Radiometric Data

Name	Diameter		Mass Moon/100	Density Water = 1
	miles	km		
Miranda	310	500	.23	3.0
Ariel	825	1330	2.13	1.3
Umbriel	690	1110	1.36	1.4
Titania	995	1600	8.05	2.7
Oberon	1010	1625	8.19	2.6

Adapted from papers by R.H.Brown, R.N.Clark, and C.Veillet, 1982–1984.

the satellites were calculated from the way their orbits precess and from the near resonances of orbital periods among the satellites.

Because of limitations in observing techniques, the diameter for Miranda given in the table could vary by as much as 120 miles (200 km) either way, the diameters of the other satellites by as much as 45 miles (70 km). The mass of Miranda was very uncertain too, as was its density. Masses of the other satellites could be off by as much as 40% for Umbriel and between 12 and 22 percent for the others, with consequent uncertainties about their true densities. These four Uranian satellites are comparable, at least in size, with the Saturnian icy satellites Dione, Iapetus, and Rhea, whose densities are between 1.2 and 1.4. But the largest of the four, Oberon and Titania, appeared to be denser than any of the Saturnian satellites, and might be expected to have larger rocky cores since the satellites are large enough to have melted and differentiated.

If these densities assumed for the Uranian satellites were real, then these satellites did not fit the pattern associated with the Jovian satellites of primordial heat from contraction of the planet determining what materials would be accreted and condensed into the satellites. The formation of these satellites of Uranus may have been associated with the catastrophic event that pushed Uranus into rotating on its side.

It is probable that Uranus condensed from the primordial nebula spinning in an upright position like the other planets. Satellites are usually thought of as originating in the same plane as the equatorial plane of the spinning planet that has condensed. It would appear, then, that the Uranian satellite system was formed after the event that caused Uranus's axis to become inclined 98 degrees to its orbit. One possibility is that the satellite system formed from the capture of a large body which subsequently disrupted into the satellites. The relatively large inclination of Miranda's orbit to the other satellite orbits (3.4 degrees) implies that there may also have been spectacular changes in the elements of the orbits of all the Uranian satellites during their evolution.

In March 1977 an opportunity occurred to observe the occultation of a 9.5-magnitude star (SAO 158687) by Uranus. This opportunity offered scientists a chance to obtain a better measurement of the diameter of Uranus by timing how long the planet, moving along its orbit, hid the star from view. Since the velocity of Uranus was known, the diameter could be calculated from the time of obscuration. Also, by use of the new photometric techniques, the change in light intensity could be measured more accurately than hitherto and was expected to provide some information about the atmosphere of the distant planet.

The occultation had been predicted well in advance—by Gordon Taylor in 1973—so groups of scientists from several countries had planned observational programs. Robert Mills of Lowell Observatory journeyed to Australia for his observation. Others went to South Africa. The faint star was located in Libra which has a southern declination and could be observed high enough in the sky to reduce Earth's atmospheric effects only in the Southern Hemisphere. Mills planned to use the 24-inch (61-cm) reflecting telescope of the Planetary Patrol Program managed from Lowell Observatory. The telescope was located near Perth in Western Australia.

James J. Elliot of Cornell University took a different approach. He planned his program to use a high flying aircraft, NASA's Kuiper Airborne Infrared Observatory, so that his observations could be made from a position above much of the distortion-producing atmosphere. He also flew south and used the 36-inch (91-cm) aperture telescope carried by the aircraft.

Because there was uncertainty about the precise time the occultation would start, the astronomers started observations several minutes in advance of the expected time. This was fortuitous because it provided data about the intensity of the starlight while the star appeared some distance from the planet.

Additionally a true occultation of the star did not take place as viewed from Perth. Instead the star grazed the edge of the disk of the planet (figure 2.3). The Lowell astronomer discovered that the light from the star mysteriously varied in intensity at a distance from the planet, and repeated the performance after the grazing incidence. Something had interrupted the light rays in their passage from the star to Earth, something in space close to Uranus. The observations from the high-flying aircraft over the southern Indian Ocean, where there was a full occultation, confirmed these anomalies. The data showed that the patterns before and after occultation were the same but reversed, like a left hand and a right hand compared. At first, scientists were skeptical about the observations being evidence of a ring system. But soon the skepticism changed to excitement as observing teams from other nations also confirmed the occultation data. Was there a swarm of small satellites circling Uranus? Nothing could be determined visually from Earth. It appeared more likely that Uranus possessed five, possibly seven, very dark and narrow rings

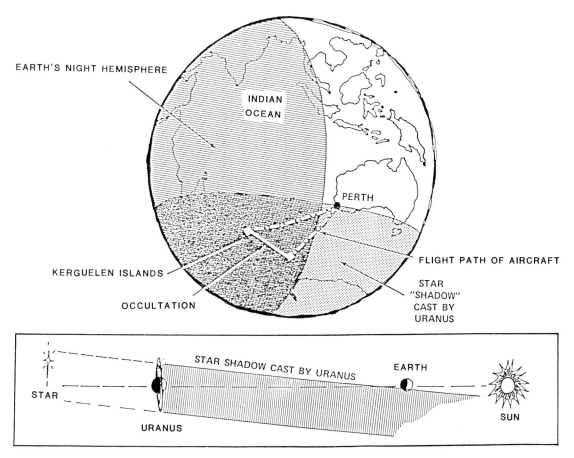

FIGURE 2.3
Diagram showing the configuration for the occultation at which the rings of Uranus were discovered. The bottom diagram shows how the shadow of Uranus and its rings, cast by the starlight, just grazed the Earth's Southern Hemisphere. (NASA drawing)

with wide gaps between them. From the light patterns observed during the occultation at least five rings were identified.

In 1978 the rings were photographed from Earth using an infrared wavelength of light that reduced the glare from Uranus. B. Thomsen used a charge-coupled device with the 200-inch (5-m) telescope at Palomar Observatory to obtain an image showing a halo produced by the brightest and outermost epsilon ring. It confirmed that the ring was elliptical and wider at one part (around apoapsis) than at the opposite part (around periapsis).

Occultations of conveniently bright stars by Uranus are infrequent and it did not seem feasible to check quickly on these first observations of the Uranian ring system. However occultations by much fainter stars were observed at other wavelengths than visual light, and they not only confirmed the presence of the rings but also led to the discovery of additional rings, and provided more

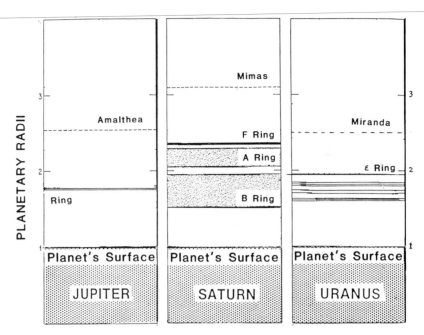

FIGURE 2.4
The ring systems of the three planets known to possess rings are here drawn to a relative scale together with the orbit of the inner-most large satellites.

information about their size and structure. By 1986 nine rings had been identi-fied and there was indication of other subtle ring structures.

All the rings appeared to be extremely narrow and sharp edged. They formed a system very different from those of Jupiter and Saturn, but also with some important similarities. If the size of the system is scaled to the size of Uranus relative to Jupiter and Saturn, the three systems are about the same size (figure 2.4). All rings are within the Roche limit, inside of which small satellites would tend to disperse into rings after collisions. However, because of the smaller gravity of Uranus, its atmosphere might extend into the ring system depending on the temperature of the very high atmosphere. If it did so, atmospheric drag might be expected to remove particles from the rings and result in the Uranian system's having a skeletal appearance compared with the Saturnian system. The Voyager imaging of the rings was expected to investigate this aspect of whether or not there is material between the thin rings identified from Earth.

The occultation observations and calculations concerning the ring system also provided information about Uranus itself. Occultation data established the equatorial radius (15,940 miles or 25,650 km at cloud tops) and polar flattening of the planet (0.024) with greater accuracy. From the polar flattening a rotation rate of 15.5 hours was calculated for Uranus, which was somewhat shorter than

TABLE 2.3
Rings of Uranus From Occultation Observations

Name	Radius		Width	
	miles	km	miles	km
6	26,014	41,864	2.5	4
5	26,266	42,270	2.5	4
4	26,470	42,598	2.5	4
alpha	27,807	44,750	4.4	7
beta	28,394	45,694	5.0	8
eta	29,334	47,207	37.3	60
gamma	29,612	47,655	2.5	4
delta	30,033	48,332	2.5	4
epsilon	31,803 mean	51,180	62.0	100
			12.4	20

previously derived results from Earth-based observations and those made from a high-flying balloon.

William Herschel had also looked for rings of Uranus, thinking that ring systems might begin with Saturn. He presented the results of his observations to the Royal Society of London in December 1797. Much earlier, in 1782, he has carefully observed Uranus to try to find out if the planet was flattened at the poles, as were Saturn and Jupiter. In February 1787 he wrote that he could not detect any ring around Uranus but that the planet did not appear quite round. He recorded early in March of that year that he saw two opposite points on the planet which he suggested might be two rings at right angles to one another. By March 8 he acknowledged that these might be caused by defects in the telescope system. In March 1789 he again saw a ring. He tried turning his mirror through 90 degrees to see if the ring moved with the turning of the mirror. It did not. He applied higher powers. Th. ring still appeared. "The ring is short," he wrote, "not like that of Saturn. . . . the two ansae seem of a colour a little inclined to red. . . . When the satellites are best in focus, the suspicion of a ring is strongest."

In February 1792 he recorded seeing the ring again and mentioned that his instrument showed excellent focus on a nearby double star. He suspected a scratch on his optical system, possibly on the mirror. So he again rotated the mirror 90 degrees. The suspected ring still stayed in place as before. It could not be a defect of the mirror. But in March 1792, with a much improved mirror, he could not see any trace of a ring of Uranus. Other observations made in 1794 and 1795 also gave negative results. Although he thought at one time that he had seen appendages to the planet he finally concluded that

Uranus did not have rings resembling those of Saturn. In his report to the Royal Society he stated "Placing therefore great confidence on the observation of March 5, 1792, supported by my late views of the planet, I venture to affirm, that it has no ring in the least resembling that, or rather those, of Saturn."

Uranus was favorably situated high in northern skies when Herschel suspected the ring system. Also, Herschel was a most astute observer; he was not troubled by city lights and he had excellent telescopes of great magnifying capability. It could be that when the rings were presented in plain view to Earth in 1775 they could not be seen, and when edge on in 1798 they could not be seen, but as has been observed with the faint crepe ring of Saturn, they could appear brighter when seen in a mid position in 1782 to 1790. This might account for the appearance and disappearances during Herschel's observations. Consequently it may have been possible for Herschel to have observed the rings while other astronomers could not do so and from then on the rings were forgotten; no one looked for them. Herschel, incidentally, had no idea that Uranus was tilted on its side.

Also there was the speculation, mentioned by the late Charles Capen in one of his papers, that a small satellite might have broken up close to Uranus relatively recently to give rise to the rings which Herschel observed 200 years ago. But since then the rings could have been losing material and becoming fainter.

The discovery of the rings of Uranus was one of the really exciting discoveries made by Earth-based astronomy, it was the first discovery of an important new feature since Pluto was found in 1930. To the date of the ring discovery in 1977, Saturn had been the only planet known to have rings and the study of those rings had posed many problems of dynamics since they had first been observed by Galileo.

Perturbations of ring particle orbits by the satellites of Saturn had long been accepted as being responsible for the features of the Saturnian rings, their boundaries, and the gaps between them. Cassini's division, for example, appeared to correspond with ring particles having one half the orbital period of the satellite Mimas. However with the advent of more powerful digital computers, models of ring systems and simulations of the evolution of rings in the presence of satellites did not fit the observed details. There were also many conflicting arguments about the size and composition of ring particles and about the thickness of the rings. No one appeared to have anticipated the discoveries of the Voyagers when they flew through the Saturnian system, that the rings would, in fact, consist of very many individual ringlets even in the spaces between rings, spaces which appear dark from Earth.

The discovery of the rings of Uranus gave theoreticians another ring system for them to analyze and to solve problems concerning ring dynamics. The later discovery, by Voyager, of a ring system around Jupiter, and even later suggestions from occultation observations of partial rings around Neptune, really stimulated an intensive interest in ring systems.

Rings around planets are important systems to study because processes that were instrumental in forming the planets and their satellites may be still working on the ring systems today. Consequently the study of ring systems offers the potential of gaining an understanding not only of how ring systems originate and evolve, but possibly of how planets and their satellites evolve.

It is not known if the systems formed at the same time as satellite systems or if they originated later from the breakup of a large body, such as a satellite, or from a slow accretion of material that should have formed a satellite, or from meteoroid bombardment of small orbiting bodies. If the Uranian rings, for example, had formed at the same time as Uranus (presumably several billion years ago) there has been time for particles with diameters exceeding a few millimeters to have diffused outward from the rings and smaller particles to have fallen to the planet. The narrowness of the rings caused astronomers to consider that the ring system must either be of recent origin or the particles making up the rings must be confined in orbit by some external force, possibly by the shepherding effects of satellites. The second alternative was preferred because there was no evidence of wide diffused rings in the Uranian system.

However, there is no simple correspondence of resonances with the known Uranian satellites that might account for longevity of the rings. The discovery of shepherding satellites at Saturn gave more credence to the idea of there being small satellites in the Uranian system that are unobservable from Earth but would act to maintain the ring system.

The particles of the rings are kept in orbit by tidal torques exerted on them by the shepherding satellites. This theory was modeled in detail by P. Goldreich and S. Tremaine in a paper published in *Nature* in 1979, and has been further refined by many subsequent studies. In the simple model, a ring particle is attracted toward the satellite as it passes it, thus changing the path of the ring particle.

Shepherding satellites are small satellites, orbiting close to or within rings, that perturb the orbits of the ring particles to prevent those particles from diffusing into or out of the rings. Figure 2.5 is a simplified diagram of how ring shepherding might work. In accordance with Kepler's law applying to bodies orbiting another larger body, the inner satellite travels faster than the innermost ring particles and overtakes them. As it does so its gravity, coupled with higher speed, acts as a slingshot on ring particles and speeds them up, just as Voyager was accelerated as it swung past Jupiter and Saturn. By contrast, the outer satellite moves slower than adjacent ring particles. Its gravity and speed difference act in the opposite way and force ring particles closer to the planet, just as the Mariner 10 spacecraft in a swingby of Venus was pushed toward the

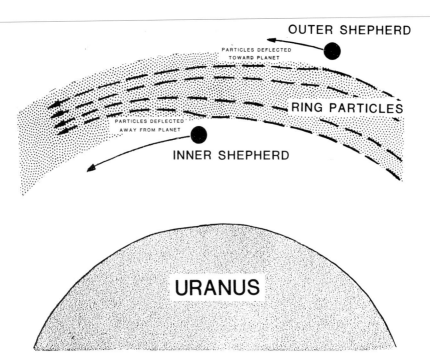

FIGURE 2.5

A simplified explanation of ring shepherding. The outer satellite travels more slowly than the ring particles. Its gravity slows the particles so that they fall inward toward the planet. The inner satellite travels faster than the ring particles. Its gravity speeds up the particles so that they move outward. Thus the narrow ring structure is maintained between the two satellites.

Sun for a rendezvous with Mercury. In this way two shepherding satellites, one outside and one inside a ring, can constrain ring particles to remain as a narrow ring.

The ring particles also lose some orbital energy by collisions with other particles between passes of the shepherding satellite. If there are two satellites at work, the ring particles move into the ring until they reach a state of equilibrium with the tendency of ring particles to spread outward and inward. This is the sharp edge of the ring. Additionally the ring particles exert torques on the satellites urging them away from the ring, unless a satellite is anchored by a resonance with other objects in the Uranian system—e.g the large satellites.

These shepherding satellites would need to have diameters of only a few kilometers to act as shepherds or guardians of the rings. If two satellites are needed for each ring, then a system of 18 such satellites could account for stability of the Uranian ring system and prevent it from dissipating. Another possibility for the rings' longevity is the action of a single large satellite or a group of very small satellites within each ring gap.

Pioneer 11 discovered a thin F-ring outside the main ring of Saturn, and the Voyagers discovered that this ring was maintained by two shepherding or guardian satellites. Also the discovery of small satellites just outside the rings of Jupiter supported this explanation.

The satellites also raise waves within the rings, as observed in the Saturnian system. Resonances amplifying these waves can also contribute to creating sharp edges to rings. This is believed to be the case for the sharp outer edges of the A and B rings of Saturn.

In 1982, following other occultations of stars by Uranus, it was found that most of the rings of Uranus are inclined to the equatorial plane of the planet by as much as several hundredths of a degree. This was surprising. One explanation was that shepherding satellites in inclined orbits prevented the ring particles from drifting into the equatorial plane as would otherwise be expected of a ring system.

So the rings of Uranus were assuming increasing importance in our studies of the evolution of the Solar System. It was hoped that Voyager's flight through the Uranian system would provide greater detail about the system, and that the spacecraft's viewpoint from beyond the planet might provide additional information about the ring structure when the ring system was viewed by forward-scattered light, something that cannot be done from Earth. Astronomers were seeking clarification of how a ball of diffuse gas and dust, the primordial nebula, could evolve into a central sun surrounded by orbiting planets, their satellites, and rings, and how other objects such as asteroids, comets, and meteoroids are formed also. Even further afield, studies of ring systems have the potential of applying similar reasoning to understanding the spiral galaxies and accretion disks on the immense scale of those galaxies.

The three other Uranian rings known before Voyager's encounter were discovered at another stellar occultation on April 10, 1978 by Eric Persson using the 100-inch (2.5-m) telescope at Las Campanas Observatory in Chile. Intended simultaneous observation by other Caltech astronomers using the Hale 200-inch (5-m) telescope at Palomar Observatory were foiled by fog.

The star occulted by Uranus was 100 times fainter than the star which was used for the initial discovery. But by using the infrared region of the spectrum, at which Uranus reflects poorly, the starlight changes could be measured more accurately without being masked by the brilliance of Uranus.

Again, the rings were clearly revealed as the star (KM5 of 10th magnitude) passed behind the Uranian system as viewed from Earth. The KM star was one of a number of faint stars catalogued by A. R. Klemola and B. G. Marsden who predicted in 1977 the occultations by Uranus of stars that might be useful for occultation research during the period from 1977 to 1980. The new occultation observations also confirmed that one of the rings was relatively broad and eccentric and precessed around the planet. Also they confirmed that the

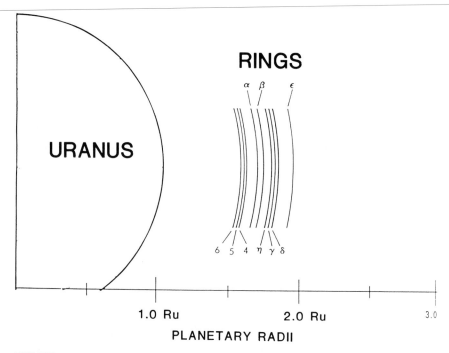

RINGS

α   β   ε

URANUS

δ 5 4 η γ δ

1.0 Ru   2.0 Ru   3.0

PLANETARY RADII

FIGURE 2.6
The nine rings of Uranus discovered from occultation data observed
at Earth are shown.

rings are complete rings and not partial arcs as had been suggested at one time
after the initial discovery.

At the oppositions of Uranus occuring in 1978 through 1983 the planet was
observed at the infrared wavelength of 2.2 micrometers. At this wavelength the
blue-green planet appears fainter than at visual wavelengths; consequently radiation from the rings is not masked by that from the planet. Infrared contour
maps showed the rings, but not individually. A striking feature was a variation
in azimuthal brightness which was attributed to the variations in width of the
epsilon ring. The surface of this ring, in fact, amounts to almost three-quarters
of the surface area of all the rings in the system.

As Voyager 2 sped toward Uranus, the features of the rings had been fairly
well established by Earth-based observations of occultations. Nine rings had
been identified (see figure 2.6). These were between 26,000 mi (41,830 km)
and 32,000 mi (51,600 km) from the planet's center. Eight were nearly circular
and less than 7 miles (11 km) wide. The outermost epsilon ring was eccentric
and varied 500 miles (800 km) in radius. Its width at periapse of 31,000 miles
(50,800 km) was 12 miles (20 km) and at the apoapse of 32,000 miles (51,600
km) was 60 miles (96 km). The epsilon ring precesses around Uranus with a
period of 264 days.

All the rings are below the orbit at which a satellite would revolve synchronously with the rotation period of the planet, and within the Roche limit. Several of the rings appeared to have some structure—i.e. separate components. The rings have slight inclinations to the equatorial plane of the planet and the inclination appeared to decrease inward toward the planet.

The most widely accepted explanation for the longevity of the rings, that of shepherding or guardian satellites, also explained the eccentricities, the inclinations, and the sharp edges to the rings. The precession of the rings was explained by their gravity.

There were also many unanswered questions about the planet Uranus itself. At an early press conference in connection with the encounter of Voyager 2 with Uranus, Bradford Smith, head of the imaging team, commented that Uranus is "a planet of unfathomable mystery."

Major questions included the period of rotation, the oblateness, the exact size, the atmosphere and its composition, circulation patterns, and clouds, the magnetic field, and the magnetosphere.

Because of the lack of surface features discernible from Earth, the rotation rate of Uranus was somewhat in doubt. Indirect evidence suggested a period of about 16 hours. A period of 10.8 hours was derived from Doppler shift spectroscoptic techniques first applied in 1912 by Lowell and Slipher. The technique relies upon observing the shift or tilt of spectral absorption lines from areas close to the approaching and receding limbs of the planet. In the 1930s other researchers obtained the same rotation period to confirm earlier estimates, which remained in vogue until the 1970s. But in 1975 R. F. Griffin and James E. Gunn obtained a rotation period approximating 17 hours using this technique on the 200-inch (5-m) Hale telescope at Mount Palomar. However, other observers using a telescope at Kitt Peak National Observatory subsequently derived a period of 24 hours using a similar technique. But spectral measurements by different observers on the same instrument led to a period of 15.5 hours.

The period can also be estimated from the oblateness—the amount of polar flattening—of a rotating planet by making assumptions about the interior, about the gravitational moment of the planet. This latter can be derived from the precession of orbits of satellites. Herschel tried to determine the oblateness of the planet by direct observational measurement. The measurement cannot be done when the poles of the planet face the Earth, so there are periods when such observational attempts would be meaningless. In 1970 Dollfus summarized the earlier measurements of polar compression as 0.033. In 1972 Robert E. Danielson established a polar flattening of 0.01 based on a computer-generated composite of images obtained with a telescope carried to high altitude by the Stratoscope balloon mission. Later the data from Stratoscope were reassessed. From it a polar flattening of 0.022 was derived. Timings of stellar occultations were also used, and they gave a value of 0.024.

As for finding the gravitational moment of Uranus, the precession of the Uranian satellites is very small and could not easily be measured accurately. However, the discovery of the rings and their rapid rate of precession permitted a gravitational moment of 0.00335 to be calculated. These values for polar flattening and gravitational moment gave a period of nearly 17 hours—quite different from some earlier estimates of the period.

Fluctuations of brightness offered another way of determining the period of the planet, and this was used starting in the late 1800s, but initially without success. A period of 10.8 hours was estimated from brightness variation observed in 1916, and as recently as 1934. However, in the 1970s, earlier data were reexamined and a period of nearly 22 hours was estimated.

The question of the rotation period was an important one that Voyager was expected to answer. If the rotation period did correspond with that calculated from polar flattening, then the interior model assumed for those calculations would be acceptable. Voyager was expected to resolve the rotation question by observation of clouds moving around the planet or by observations of cyclical changes in the magnetic field as the planet rotates.

An important question is how the interior of Uranus might differ from those of Saturn and Jupiter. Uranus is smaller and denser than either of these two planets, which would suggest that it has lost some of its gaseous envelope or was not able to collect as many light elements as did the others. Pressures within Uranus are much less than those within Jupiter and Saturn. They would be too low to account for a density of 1.6 without there being a rocky core and a deep shell of water, methane, and ammonia.

The rotation rate and the gravitational moment allows the development of models of what the interior might be. The most plausible suggestion is that the planet consists of a central rocky core of iron and magnesium silicates surrounded by a deep ocean of water, methane, and ammonia, topped by a deep atmosphere of hydrogen, helium and methane (figure 2.7). The rocky core of Uranus is estimated as being 9,300 miles (15,000 km) in diameter. The water/ice ocean is about 6,500 miles (10,500 km) deep, and the hydrogen/helium atmosphere is about 5,200 miles (8,400 km) deep. A comparison with the interiors of Jupiter and Saturn is shown in figure 2.8.

There are major questions concerning the chemical composition of Uranus compared with the abundances of elements in the Sun and the other giant planets. This is important to gaining an understanding of how the outer planets may have formed from the primordial nebula, and the composition of that nebula. Also it is important to know the chemical composition of the atmosphere of the planet, because this gives us clues as to the interior and the

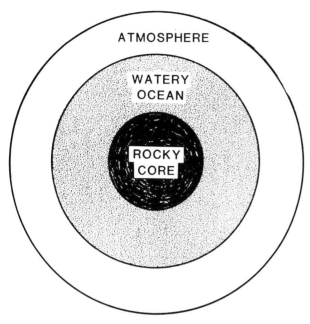

FIGURE 2.7
A possible structure of Uranus showing its inner core, deep ocean, and atmosphere.

evolution of the atmosphere. The interior of the planet may be differentiated. We would like to know how. Does it indeed have a rocky core with a definite boundary surface between the core and the possible deep ocean? Does this deep ocean have a clear boundary with the atmosphere?

The bulk composition of Uranus has to correspond to the planet's mean density, our concept of which changed dramatically from the 1960s to the 1980s as more accurate measurements of the size of the planet were made. The mean density of Uranus accepted in the 1960s was 1.36. This had been amended downward to 1.31 by 1975 as the diameter measurements increased. The mass of Uranus is about 14.5 times that of Earth.

The internal structure of Uranus could be more accurately inferred when Voyager provided more precise data to be used to determine the harmonics of the Uranian gravity potential, the diameter and polar flattening of Uranus, and the planet's period of rotation. It was estimated that Voyager would provide gravitational harmonic data no more precise than that already obtained from Earth-based observations, but that data on the rings would improve the calculation of the harmonics. The radio occultation would, however, provide a more accurate determination of the diameter of the planet. If clouds can be imaged, or an offset planetary magnetic field can be detected, the rotation period would be established with greater accuracy.

While the Earth's atmosphere may be likened to the fuzz on a peach, the atmosphere of Uranus may be likened to the skin of an aorange. The atmos-

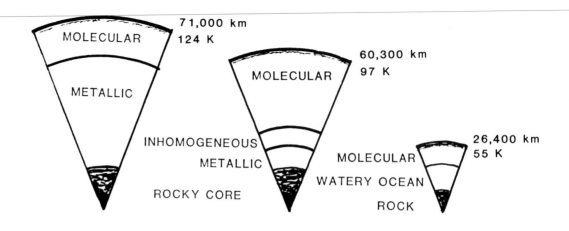

FIGURE 2.8
The inner structure of Uranus is here compared with those postulated for Jupiter and Saturn.

phere was believed to represent 19% of the mass of the planet, the core about 43%, and the icy water mantle about 38%. In 1952 Gerhard Herzberg identified hydrogen on Uranus, but its abundance was uncertain. The atmosphere was accepted as most probably being mainly hydrogen mixed with helium, but the percentage of helium was very much in doubt. Some deductions from Earth-based observations suggested a helium constituent amounting to 40% of the Uranian atmosphere, which seemed abnormally high.

Although absorption bands had been observed in the spectrum of Uranus as far back as 1869, confirmation that they arose from absorption by methane gas did not come until 1934, when V. M. Slipher of Lowell Observatory obtained spectra of Uranus out to 8500 angstroms in the near infrared and, working with A. Adel, matched the spectra with laboratory spectra of methane.

In 1967 T. Owen confirmed the presence of weak methane bands in the spectrum of Uranus, adding another known constituent. Methane absorption between 9600 and 10,000 angstroms is stronger in the spectrum of Uranus than in those of Jupiter and Saturn. Ammonia is believed to be present. Although ammonia bands are clearly identified in the spectrum of Jupiter, they were not found in the Uranian spectrum. This is probably because the lower temperature of Uranus compared with that of Jupiter causes the clouds of ammonia to condense deeper in the Uranian atmosphere, where they cannot be detected spectroscopically (see figure 2.9).

However, microwave radiation can be analyzed to show ammonia abundances. Microwaves at a wavelength of centimeters show less ammonia than expected if it were present in the same abundance as on the Sun. This might be explained by ammonia condensing into clouds. Strangely the centimeter radiation from Uranus has changed considerably over time. This is probably a result of the planet's current presentation of polar regions rather than equatorial regions to the Sun. The polar regions may lose ammonia to clouds more than do the equatorial regions.

There is no direct evidence of the abundance of water vapor in the atmosphere. Again this may be because water is confined below a level at which the vapor condenses into clouds while the upper, observable atmosphere is depleted of water vapor.

Concerning the temperature and pressures at levels in the atmosphere of Uranus, no phases can be observed from Earth and the bolometric albedo was not measurable but was assumed to be about 0.35. Bolometric refers to the total reflectivity of the planet over all wavelengths, and not just to the reflectivity at wavelengths of visible light. The planet is very reflective in blue and green with great absorption in red and near infrared.

The planet experiences a long year of 84 Earth years and long seasons as first one pole faces the Sun and then the other. Large seasonal changes of heat input to the atmosphere would be expected. Poles faced the Sun around the years 1775(N), 1817(S), 1859(N), 1901(S), 1943(N), 1985(S), and the planet's equatorial regions faced the Sun in the years 1756, 1798, 1840, 1882, 1924, and 1966. If Earth's axis were tilted like that of Uranus, the weeks-long, or months-long, winters and summers experienced in Arctic and Antarctic regions would extend to our planet's equatorial regions. Note that in early records the poles of Uranus are often oppositely labeled; the north pole according to current nomenclature of the International Astronomical Union being referred to then as the south pole. Uranus is at perihelion (closest to the Sun) and consequently closest to the Earth when at opposition at times when the planet's equator is presented to us. Perihelion was in 1798, 1882, and 1966. The equator is also presented Earthward at aphelion.

The equilibrium temperature at Uranus' distance from the Sun is $-357°$ F ($-216°$ C) (based on a bolometric albedo of about 0.35 and with the equatorial regions toward the Sun) compared with $-1.4°$ F ($-17°$ C) for Earth. With a pole facing the Sun, the temperature might be expected to rise about 10 degrees, since the planet would be heated in the one hemisphere facing the Sun. The other hemisphere, could its temperature be measured, would be expected to be 10 degrees cooler.

In the 1980s a whole-disk Uranus temperature of about $-337°$ F ($-205°$ C) was measured. At the cloud tops, however, temperatures were warmer, about $-256°$ C.), which is about 125° F. (70° C) colder than the coldest nighttime temperature in Earth's polar regions. These low cloud-top temperatures corre-

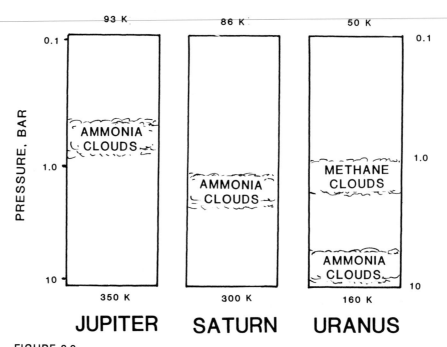

FIGURE 2.9
Comparisons of the outer atmospheres of Jupiter, Saturn, and Uranus illustrate the levels at which methane clouds are believed to form. The clouds of Uranus and Saturn are deeper in the atmosphere than those of Jupiter and are more difficult to observe.

spond to the temperature range in which methane, acetylene, and ethane turn to clouds.

Uranus is peculiarly different from the other outer planets in that it does not appear to emit more heat than it receives from the Sun. Because Uranus is smaller than Jupiter and Saturn, the planet's lack of internal heat might be easily explained on the basis of size and different internal structure. The mystery then becomes why Neptune, which is about the same size as Uranus, should be different.

Radio measurements gave even higher temperatures. Radio emission from the planet is in the range 3 mm to 11 cm, and from observations of these emissions a temperature of $-270°$ to $-78°$ F ($-168°$ to $-61°$ C) was deduced. However, the temperature depends upon the size attributed to the disk—i.e. to the radius of the planet, which was rather uncertain. Until fairly recently the radio measurements could not resolve areas of the disk, so derived temperatures referred to the whole disk of the planet. More precise measurements in the 1980s increased the accepted angular size of Uranus and reduced the value accepted for the radio brightness temperature. This radio brightness temperature exceeded the equilibrium temperature—temperature achieved by balancing incoming and reradiated energy from the Sun—expected at the planet, possibly because of an effect of ammonia saturated vapor. In fact, at centimeter wavelengths Uranus appears to have a temperature exceeding that at Jupiter and Saturn as measured at the same wavelengths. This could be because the atmosphere of Uranus is not as opaque as those of Saturn and Jupiter and permits the recording at Earth of radio emissions from deeper and warmer levels of the atmosphere.

High resolution radio temperature maps of Uranus made in the early 1980s by means of NASA's infrared telescope facility and the very large array telescope of the National Radio Astronomy Observatory displayed some assymmetry across the face of the planet. The south pole was facing Earthward at that time, and the brightest part of the disk was toward the polar region. Moreover, other work over the years has indicated that Uranus was cooler in the 1960s and warmed up as the pole faced more and more toward the Sun in the late 1970s. This radio brightening effect was observed over a wide range of wavelengths from millimeters to centimeters. In the mid 1960s the brightness temperature of Uranus in the 10 to 21 cm range was around $-189°$ F ($-123°$ C), while in the early 1980s it was $-9°$ F ($-23°$ C). In the millimeter wavelength range, the temperature change was not so pronounced. In both cases the temperature appeared to reach a plateau a few years before the date the pole was aligned toward the Sun. The cause may be that the atmosphere is more opaque at the equator than at the poles so that warmer lower levels are observed at the polar regions. The reason, however, may not be clearly stated until we have observed a whole 84-year cycle of radio emissions from Uranus.

The question of circulation in the atmosphere of Uranus led to much speculation arising from the unusual orientation of the axis of rotation. Nothing was definite about the dynamics of the Uranian atmosphere because no motions had been observed clearly from Earth. Some infrared images suggested shading on the planet, and images obtained with charge-coupled devices revealed a dark polar hood.

When the equatorial regions of the planet are presented toward Earth, some astronomers reported seeing a bright equatorial zone with darker belts on either side of it. However, most observers of the belts reported that these were strangely inclined to the plane of the satellite orbits, some by as much as 45 degrees. However, combinations of photo images of the planet failed to confirm any definite zonal markings of that type. The disk of Uranus is virtually featureless in visible light, much like that of Venus. The atmospheric circulation of Uranus could very well be similar to that of Jupiter and Saturn, but whereas the Jovian and Saturnian atmospheres contain gaseous materials that condense into clouds of distinct colors and are banded into zonal patterns clearly visible from Earth, the components of the Uranian atmosphere are quite different. The cloud materials might be homogeneous and not develop a banded structure, or, more likely, they probably condense so deep in the atmosphere as to be hidden from observations in visible light.

The most opportune periods to look for a banded cloud structure of Uranus is when the planet presents its equatorial regions toward the Sun (and hence toward Earth) and is at perihelion, as was the case in 1798, 1882, and 1966.

Observations by Bradford A. Smith in 1976 with a charge-coupled device detector operating at the band of methane absorption in the near infrared, showed what appeared to be a dark area between equator and pole, and some limb darkening. A few years later the technique revealed a banded structure on Neptune, but nothing similar on Uranus. In 1983, images of Uranus were obtained by Smith and R. J. Terrile using a coronagraph attached to the 100-inch (2.5-m) telescope at Las Campanas observatory in Chile. While the atmospheric markings on Neptune were much the same as at the earlier period, the markings on Uranus had changed. There was not as much limb darkening and the disk appeared quite featureless.

Figure 2.10 shows an expected structure of the atmosphere of Uranus, characterized by high hazes and methane clouds hiding lower clouds of acetylene, ethane, ammonia, and water ice, with various snows falling between some of the layers. The clouds of methane and methane hazes absorb red light and give the planet its pale greenish-blue color. These clouds also hide the colored clouds at greater depths and give the planet its bland appearance.

What had stellar occultations and observations of radio emission added to this idea of the structure? The stratosphere of Uranus is much colder than those of the other giant planets. The microwave emissions from the region of the atmosphere where pressures exceed 10 bars are much the same for Uranus and Neptune, although it might be expected that because of a higher heat flow from the interior of Neptune than from the interior of Uranus, conditions would be different.

The big questions outstanding concerned finding an explanation of why Uranus has so much less cloud activity than Neptune and, of course, Saturn and Jupiter. Also, how does the orientation of Uranus affect circulation in the atmosphere during the various long seasons experienced by the planet? What causes the variations in the microwave spectrum between 1 and 21 cm? Is it depletion of ammonia? Is it a change in state leading to a pseudo-surface at a level in the atmosphere where the temperature is about 45° F (7° C), or is hydrogen sulfide at work removing ammonia from the atmosphere at a level where the temperature is about −190° F (−123° C)?

As Voyager approached Uranus scientists hoped that the various experiments planned would answer some of these puzzling questions and would settle the intriguing mystery of whether or not Uranus possessed a magnetic field and a magnetosphere.

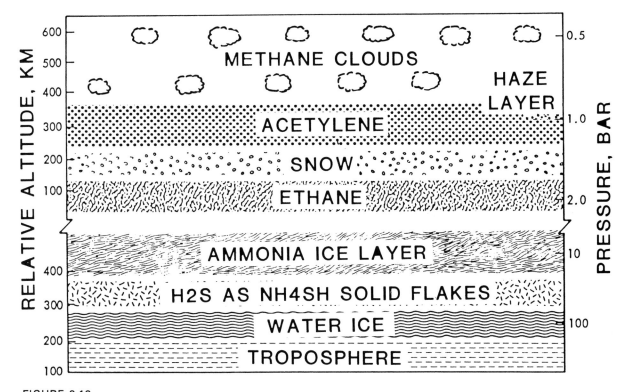

FIGURE 2.10

The structure of the atmosphere of Uranus below the methane clouds may never be determined until probes can be dispatched to penetrate the atmosphere. However, below clouds of methane we would expect there to be layers of acetylene and ethane, of ammonia ice, and of water ice. (NASA drawing)

In 1976 an IMP Earth-orbiting satellite detected a few radio signals from the direction of Uranus which might have been from radiation belts of that planet. Six years later, the International Ultraviolet Explorer satellite recorded hydrogen Lyman-alpha emissions from Uranus during the period March 1982 through September 1983. These emissions showed great variations over relatively short periods. Lyman alpha ultraviolet radiation at 1275.7 angstroms is generated when electrons return to a less excited state in hydrogen atoms. The radiation can arise when atoms of hydrogen absorb and then re-emit solar radiation (resonant scattering) or from reflection of solar radiation by atmospheric particles (Rayleigh scattering) by a layer of atomic hydrogen in the upper atmosphere, or from the impact of charged particles precipitating into the upper atmosphere from a magnetosphere as occurs in the generation of aurorae at Earth. Generation of Lyman alpha by auroral processes was observed at Jupiter and Saturn, both of which have magnetic fields.

It seemed unlikely that there could be sufficient atomic hydrogen above the methane clouds of Uranus to account by resonant scattering for the radiation observed. Rayleigh scattering was also thought to be insufficient to account for the intensity of the Lyman alpha emissions observed by the Explorer satellite. Auroral generation appeared to be the most likely source, and this would require the presence of a Uranian magnetosphere and hence a magnetic field generated by the planet. The question then was; why did the emission vary so much? One possibility was that the magnetosphere varied considerably under the influence of the changing intensity of the solar wind.

A magnetosphere of a planet is a region around the planet where the magnetic field of the planet is the dominant force controlling the movement of charged particles such as electrons and ions. For a planet possessing an atmosphere, like Earth and Jupiter, for example, the inner boundary of the magnetosphere is the ionosphere, the part of the upper atmosphere which contains charged particles. The outer boundary of the magnetosphere is called the magnetopause. This boundary is shaped by the solar wind, blunt-nosed on the side of the planet facing the Sun and streaming out away from the Sun as a magnetic tail; the magnetotail. On the sunward side the magnetopause is pushed toward a planet when the solar wind is strong, and expands away from the planet when the solar wind weakens. As the solar wind encounters the magnetosphere on the sunward side, the wind is abruptly halted or deflected around the magnetopause and produces a discontinuity called the bow shock. At the shock the supersonic solar wind blizzard of electrons and charged particles—mainly protons—is slowed down so that it can change its path and flow around the obstacle presented by the planet's magnetosphere.

The solar wind itself carries a magnetic field which interacts with the field of the magnetosphere. One effect is to link with magnetic field lines from the planet and change them from closed to open lines; in other words, instead of a field line returning to the opposite magnetic pole of the planet it is "pulled" by the interplanetary field so that the field line is connected to the planet at one end only. This allows interplanetary charged particles to reach the planet, and particles within the magnetosphere to escape into space and join the solar wind. The open field lines ultimately connect with other planetary field lines downstream in the magnetic tail. Charged particles become trapped within the magnetosphere spiraling in north-south oscillations along field lines and give rise to radiation belts first discovered in Earth's magnetosphere by instruments carried in Earth-orbiting satellites.

The magnetosphere also contains important electric fields as well as the magnetic field. The electric field becomes very strong where the magnetic lines rejoin in the magnetotail and there accelerate charged particles to high velocities.

The plasma of energetic particles and ions in a magnetosphere is maintained by the solar wind and the ionosphere of the planet and by satellites orbiting within the magnetospheree. The magnetosphere also loses particles and ions to satellites, to the planetary atmosphere, and into interplanetary space. So it is a very dynamic region. Toruses of neutral gases are formed within some magnetospheres e.g. a torus of sodium and potassium atoms from Jupiter's satellite Io, from which charged particles are generated.

Trapped particles in magnetospheres generate radio waves. Radio waves from Jupiter were first detected by K. L. Franklin and B. F. Burke in 1955. Radio waves at decimetric wavelengths were later identified as originating from electrons spiraling along magnetic field lines, as had been recognized as taking place in Earth's magnetosphere by observations from satellites. That Jupiter had a magnetic field was thus first established before spacecraft visited the planet to confirm the presence of such a field. Unfortunately, no such radio waves from Uranus could be detected with certainty, but neither could they be detected as emanating from Saturn. Subsequently, spacecraft flying by Saturn confirmed that it, too, has a magnetic field.

Since Uranus is a large planet spinning relatively rapidly on its axis, scientists regarded the auroral explanation of the Lyman alpha radiation as probably valid, and they accepted that Uranus most likely possessed a magnetic field. If so, a Uranian magnetosphere would be expected to have an unusual shape and physical properties (figure 2.11). The orientation of the planetary axis—assuming that the magnetic field axis, as with other planets, was fairly close to the axis of rotation—leads to big changes with respect to the direction of the solar wind. When the planetary axis points toward the Sun the solar wind will plunge down close to the surface along the magnetic axis through a kind of funnel. Aurorae would be expected to concentrate in a spot near the pole rather than as an auroral ring as observed on Earth. Additionally, at this time the tail current sheet would be expected to be a cylinder rather than the plane experienced in Earth's magnetosphere. The rotation of the planet, directed along the tail, would also tend to produce a twisted, helical field. Only when equatorial regions faced the Sun would the geometry of the magnetosphere be expected to be similar to the magnetospheres of the Earth, Jupiter, and Saturn.

If such a Uranian magnetic field existed, it would be interesting to ascertain whether the satellites were immersed in it and what interactions took place between the satellites and the magnetosphere. Would the satellites be sources or sinks for the energetic particles of the magnetosphere? Because the satellites are relatively small and unlikely to possess atmospheres they would be expected to act as sinks, sweeping up energetic particles along their orbits. The Saturnian satellites Mimas, Enceladus, and Tethys sweep regions of the Saturnian magnetosphere clear of particles. In fact, a new satellite of Saturn inside the orbit of Mimas was first detected by its sweeping effects on the magnetosphere of the planet. As at Saturn, the Uranian rings would be expected to affect the inner magnetosphere. Saturn's rings sweep particles from the inner magnetosphere to keep it virtually free.

A major question about Uranus which cannot be easily resolved even by

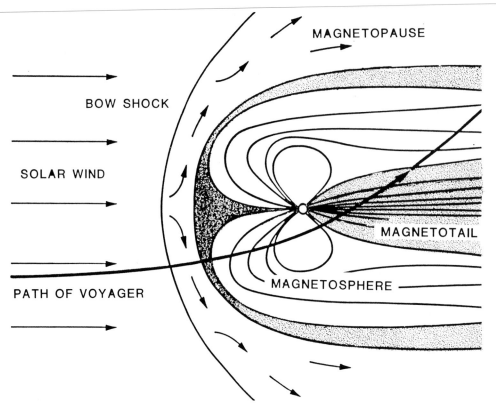

MAGNETOPAUSE

BOW SHOCK

SOLAR WIND

PATH OF VOYAGER

MAGNETOTAIL

MAGNETOSPHERE

FIGURE 2.11
The magnetosphere of Uranus was expected to be unusual because of the inclination of the planet's axis of rotation. At the time of encounter of Voyager with the planet, the magnetic field lines were expected to be in this configuration with the pole of the planet facing the Sun and a circular helical plasma sheet stretching out in the planet's magnetotail. (NASA drawing)

flyby spacecraft is why its rotation axis is so different from the other planets. What could have caused this axial tilt? One possibility is that soon after the planet's formation it was hit by an Earth-sized body. An impacting speed of about 40,000 mph (64,000 kph) might be sufficient, according to some calculations, to push the planet onto its side. But then there is the question of the satellites. It is unlikely that they were formed before the planet was pushed onto its side, since their orbits are all so close to the equatorial plane of the planet. More likely they were formed after the catastrophic event, perhaps from debris remaining after that event. On the other hand, our ideas of how planets form might be fallacious. Planets might not have formed from a nebula in a disk-like configuration.

There are other anomalous situations in the Solar System, such as the slow retrograde rotation of Venus. That planet might have flipped completely over since its formation. There are also peculiarities of Triton and Neptune, of Pluto and Charon which will be discussed in the final chapter. Why, for example, does Neptune, alone of the outer planets, not have a system of regular satellites? And why does Uranus not have retrograde satellites?

Perhaps some of the anomalies of Uranus and Neptune were caused by the formation process having taken considerably longer in the outer Solar System than it did in the inner Solar System. However, there is no firm theoretical reason why this time difference should be so, especially if the relative velocities of planetesimals were not high. Uranus may have been formed from condensed planetesimals which later captured gases such as hydrogen, water, ammonia, and methane from the remaining solar nebula. Planetesimals are hypothetical bodies about a mile or so in diameter which are thought to have accreted from clumps of solid material that condensed from a cooling primordial nebula from which the Solar System formed. Theory has it that the planets ultimately formed from these planetesimals whose composition varied in different regions of the solar nebula.

Another important question is why the heat flow from the interior of Uranus is much less than that from Neptune and from Saturn and Jupiter. Uranus' heat flow is less than one-quarter that of Neptune, for example.

The amount of ammonia in the atmosphere of Uranus is also strikingly different from Saturn and Jupiter. The ammonia cannot be detected in the visible spectrum, but it can at microwaves. On Jupiter and Saturn the abundance of ammonia is close to that found in solar material. For Uranus the abundance is much smaller. There appears to be a depletion of ammonia in the Uranian atmosphere. One possibility that was discussed earlier is that it is removed as a gas from the atmosphere by reacting with hydrogen sulfide and condensing into clouds, but this removal mechanism seemed to differ between equator and poles to account for the variability of centimeter radiation from the planet as different regions (pole and equator) are presented toward the Sun, and hence the Earth.

The flyby of Voyager would provide some important information to help solve the problems associated with understanding this mysterious planet. First the spacecraft images were expected to determine the rotation rate of the planet which even in the last few years had been stated from Earth-based observations as either 16 hours or 24 hours—quite a large discrepancy.

Voyager was also expected to show how pressure and temperature vary with depth in the outer layers of the atmosphere and perhaps determine the composition of the atmosphere and how much sunlight is absorbed by the atmosphere compared with energy rising from the interior. The location of clouds and hazes might be mapped and wind speeds measured at various latitudes on the planet. Confirmation of auroral activity would be sought.

The rings would be examined in detail to ascertain, if possible, the size, composition and distribution of particles forming them. Other rings would be searched for, as also would be shepherding satellites which had been proposed as the origin of forces to prevent rings from dissipating. The sizes and shapes of the major satellites of Uranus would be measured, and other smaller satellites searched for. The surface features of the satellites would be mapped to try to understand how these satellites had formed and evolved.

The strength, orientation, and charged particle population of the magnetic field would be determined if a field did, in fact, exist as postulated from the Lyman alpha emissions observed by the International Ultraviolet Explorer satellite. The inclination and the rotation rate of the magnetic field would be determined, the latter to ascertain the rotation rate of the interior of the planet where the field is generated. The orientation of the field would also be measured to ascertain where in the planet it is generated.

We read in the previous chapter how successful the Voyager spacecraft was in its encounter with Uranus. The next chapters discuss what it discovered about this enigmatic and distant world.

3

CLOSE-UP OF
A DISTANT GIANT

As Voyager approached Uranus scientists inspected the data being returned from the particles and fields experiments and radio experiments to see if there was any evidence of a magnetic field. The plasma wave equipment (PWS) detects radio frequencies in the low range from 10 Hz to 56 kHz. The first indication of sporadic radio emissions came on January 19, 1986, when the spacecraft was about 4.4 million miles (7 million km) from the planet. This appeared to be the first sign of the presence of a Uranian magnetosphere. A little later the Planetary Radio Astronomy (PRA) data started to show nonthermal radio emissions from the planet about 5 days before closest approach, when Voyager was about 3.7 million miles (6 million km) from Uranus. Polarization of the radio waves suggested that they were being generated by electrons spiraling along magnetic field lines. Periodic fluctuation cycles of about 17¼ hours became evident as Voyager approached the planet. These presumably arose from a rotating magnetic field, and gave a first indication that the rotation period of Uranus is about 17¼ hours.

The low energy charged particle instrument (LECP) which detects ions with energies less than 28 keV and electrons of less than 22 keV began registering an increase in the number of ions on January 21, 1986, three days before closest approach to Uranus, when the spacecraft was 2.2 million miles (3.6 million km) from the planet. The activity increased to a large peak followed by crossing of a bow shock, indicating the presence of a large magnetosphere.

The PWS instrument detected waves upstream of the bow shock early on 23 January, presumably caused by electrons escaping from the bow shock into the solar wind, as occurs at Earth, Jupiter, and Saturn. These waves continued to be observed until the crossing of the shock, the following day.

As Voyager continued its approach to Uranus, details of the magnetosphere and the planet's magnetic field increased. Instruments registered the crossing of the bow shock at 374,000 miles (601,000 km), 23.5 Uranian radii from the planet, on the morning of January 24. The PWS instrument showed relatively low levels of plasma waves until the spacecraft entered the inner regions of the magnetosphere about 8 Uranian radii from the planet. There, whistler-mode emissions, and bursts of unusual emissions, including an intense emission as the spacecraft passed through the plane of the rings, were recorded. Later still, as Voyager left Uranus and could obtain data within the magnetotail, the plasma waves there were of low intensity only.

Uranus has been found to have an extensive corona of uncharged hydrogen revealed in the ultraviolet observations of the planet by the ultraviolet spectrometer (UVS). Hydrogen atoms are stripped of electrons and the resultant protons become an important constituent of the magnetospheric plasma. The hydrogen corona rotates with the planet and the protons acquire this rotational energy. As the protons migrate inward or outward with respect to the planet, their energy changes. Moving inward they form the "warm" component of the magnetospheric plasma. The "hot" component results from protons formed farther out and moving a greater distance inward. However other mechanisms are probably at work also, among them a precipitation of ions into the ionosphere from the magnetosphere, giving rise to high-energy electrons which drag ions along into the "hot" region of the magnetosphere.

The energies of escape of hydrogen atoms from the big planets are 19 eV for Jupiter, 6.3 eV for Saturn, and only 2.5 eV for Uranus. Hydrogen can attain the escape velocity at Uranus and accordingly can blow off the planet and escape into space.

The PLS instrument mapped the low-energy positive ions and electrons within the magnetosphere and discovered that the plasma there has the different components mentioned above. In the inner magnetosphere the "warm" component in terms of energy has a temperature of about 10 eV, the "hot" component has a temperature of 0.7 to 3 keV, and the "suprathermal" component has a temperature between 50 and 100 eV and is connected with the warm component plasma protons. The positive ions in the plasma appear to be protons that are corotating with the planet. These are most probably derived from the hydrogen corona which the ultraviolet experiment detected surrounding the planet. Some protons may, however, originate from the solar wind.

The magnetospheric plasma consists of "hot" ions outside Miranda's orbit with a temperature of about 10 million° K. The density of charged particles there is about 0.1 ions per cc. Throughout the magnetosphere there are warm ions everywhere with a temperature of about 100,000° K and a density of approximately 1 ion per cc. These ions are mostly protons.

Plasma originates not from the solar wind, but possibly from the outer satellites or from a neutral hydrogen cloud surrounding the planet.

Within the magnetosphere warm protons rotate with the planet, and the planetary magnetic field is not modified by the plasma. The hot protons move slowly toward the planet and are absorbed by Miranda.

Corotation of the plasma with the smaller Uranus, which is smaller than and also rotates more slowly than either Jupiter or Saturn, results in much lower velocities of ions than in the inner magnetospheres of Jupiter and Saturn which, in turn, does not cause the ions to be controlled so strongly. Convection of plasma takes place in the Uranian magnetosphere. Additionally, the alignment of the rotation axis toward the Sun permits the solar wind to drive convection processes within Uranus' magnetosphere. These two effects may account for the absence of heavy ions. Were such ions sputtered from the surfaces of the satellites they might have been convected out of the magnetosphere faster than they are replenished from the surfaces of the satellites.

The cosmic ray subsystem also measured energetic protons and electrons in the magnetosphere, and detected the sweeping effects of satellites. Energetic electrons appear to be originating from the outer magnetosphere or the magnetotail and are gradually diffusing inward toward the planet to create an intense high-energy radiation region close to the planet within the orbit of Miranda. The three inner satellites, Miranda, Ariel, and Umbriel, appear to play a significant role in sweeping electrons and limiting the electron flux within the magnetosphere, much more so than the inner Saturnian satellites do at Saturn. Importantly, the sweeping action of the satellites was used to determine the shape of the magnetic field which confirmed the geometry derived from the magnetometer data.

Rapidly rising energy of electrons as the spacecraft made its close approach to the planet indicate that even closer to the planet there should be a region of intense radiation belts. The radiation belts of Uranus appear to be more intense than those of Saturn; they are comparable to the intensity of the Van Allen belts at Earth, but are, of course, much more extensive. Ions at higher energies in the belts are almost exclusively protons, quite different from the other planets, and plasma temperatures exceed 500 million° K, five times that of Earth's radiation belts. Ion bombardment from these intense radiation belts can affect the color of satellites and ring surfaces. The intensity of the proton component is sufficiently high to polymerize methane ice and, more slowly, carbon monoxide ice, and thereby darken the surfaces of the satellites that orbit within the magnetosphere. While the plasma of Uranus' magnetosphere consists mainly of electrons and protons, there are a few ionized hydrogen molecules.

When Voyager was leaving Uranus on the farside of the planet from the Sun—i.e., the night hemisphere—instruments detected decreases in the fluxes of ions and electrons. These indicated that the spacecraft was flying through a sheet of plasma connected with the magnetotail of Uranus. This was expected to be a flat sheet extending along the center of the cylindrical magnetic tail. Because the magnetic dipole of Uranus rotates askew relative to the planet's rotation, the spacecraft as it moved along its trajectory traced a spiral path relative to the magnetotail and moved in and out of the plasma sheet for about one day. Beyond that the position of the plasma sheet became variable, probably because of some effect not clearly defined at present. In the magnetotail electrons from the solar wind may acquire energy and become the "hot" electrons detected in the inner magnetosphere on the nightside of the planet.

The magnetic field of Uranus was ascertained to be proportionately the same intensity as that of Saturn. Just before the time of Voyager's closest approach to Uranus the magnetometer measured the intensity of the magnetic field as 0.00413 gauss. The great surprise was concerning the orientation of the magnetic dipole. It differed from the planet's axis of rotation by 60 degrees, much more than on any other planet explored so far. As a result the intensity of the field at the visible surface of the planet (the clouds) varies from 0.1 on the sunlit hemisphere to 1.1 gauss on the night hemisphere compared with 4.2 gauss at Jupiter's equatorial cloud tops, which is more than ten times stronger than Earth's field at the surface (0.31 gauss). The strength of Saturn's field at the equatorial cloudtops is 0.2 gauss. Variations of Jupiter's field, north to south, are 14 and 11 gauss, and Saturn's field, north to south, are 0.63 and 0.48 gauss respectively. Earth's polar field is 0.6 gauss (See table 3.1).

Because, at the time of Voyager's encounter with the planet, Uranus was spinning on an axis that was closely aligned to the direction of the solar wind flow at this section of its orbit, the magnetosphere did not wobble as it does at Earth and Jupiter. Instead it rotated in space along a line corresponding to the radius vector from the planet to the Sun (figure 3.1).

The magnetosphere extended to 17.8 Uranian radii facing the Sun and to

TABLE 3.1
Comparisons of Magnetic Fields

Planet	Dipole Moment Gauss cm³	Tilt and Polarity	Equatorial Magnetic field gauss	Size of Magnetosphere Planet Rad.
Mercury	5×10^{22}	+14	.0033	1.4
Earth	8×10^{25}	+11.7	.31	10.4
Jupiter	1.6×10^{30}	−9.6	4.28	65
Saturn	4.7×10^{28}	−0	.21	20
Uranus	4×10^{27}	−60	.25	18

Dipole moment and dipole equatorial magnetic field are analogous to total mass of a planet and its surface gravity force.

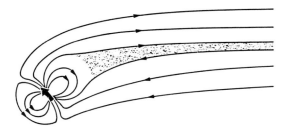

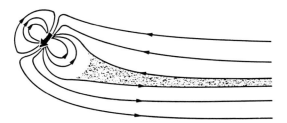

FIGURE 3.1
The offset and inclination of the magnetic field of Uranus to the spin axis of the planet causes the magnetosphere to writhe around the planet. Here it is shown at two extremes during this configuration relative to the solar wind.

about 30 at the sides. Thus all the satellites are within the magnetosphere when a pole of Uranus is toward the Sun and for much of the Uranian year. When the equator is toward the Sun the outermost satellite, Oberon, will be out of the magnetosphere on the part of its orbit close to the Sun–planet line, but at times the satellite may be within the bow shock depending upon the intensity of the solar wind. Because of the tilt of Uranus' magnetic axis relative to the equatorial plane, and hence to the orbits of the satellites, the satellites pass through great variations in the intensity and direction of the magnetic field as they travel around Uranus. The change in latitude of the magnetic field during these orbits means that the satellites can effectively sweep trapped particles over a wide path (figure 3.2). They become voracious scavengers of Uranus' radiation belts.

Satellite sweeping of the magnetospheric plasma is different at different energy levels of the particles. Those particles having an energy such that they drift around the planet at about the same velocity as the satellites travel in orbit are least affected by sweeping. Although one would expect electrons and ions to be lost from the magnetosphere by satellite sweeping at about the same rates, energetic ions appear to be removed selectively from the radiation belts of Uranus by some additional mechanism.

The rotation rate of the magnetic field is assumed to be that of the interior of the planet. The magnetic instrument data indicate a rotation rate of 17.29 hours which is considerably more than the 15.57 hours accepted before the mission. There were also indications of a strong quadrupole and of other magnetic harmonics in addition to the dipole field. Jupiter's field possesses strong quadrupole harmonics, but Saturn's does not. It has been suggested that Saturn's lack of a strong quadrupole field may indicate that the magnetic field

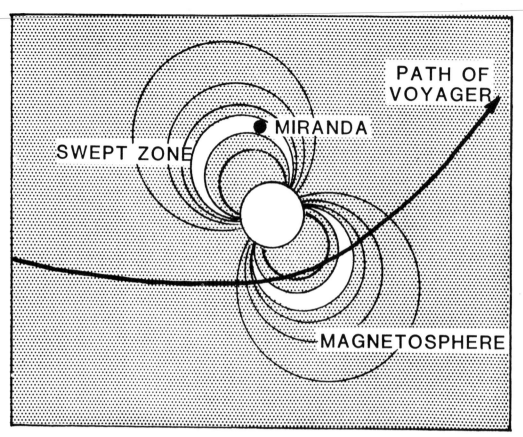

FIGURE 3.2
The inner satellites of Uranus sweep particles from the magnetosphere in wide lanes because of the offset of the magnetic axis from the planet's axis of rotation and the resultant wobbling of the field.

is generated much deeper in the planet, possibly in a smaller core of metallic hydrogen. Uranus' strong quadrupole may indicate that its field is generated in the water mantle surrounding the rocky core.

The dipole field is not only tilted 60 degrees but also offset by 4,800 miles (7,680 km) or 0.3 Uranian radii from the center of the planet (figure 3.3). The field of Jupiter is also tilted and offset; the tilt is 10 degrees and the offset is 6,200 miles (10,000 km) or 0.14 Jovian radii. Saturn's field is aligned very closely along the axis of rotation but it, too, is displaced along the axis of rotation by 1490 miles (2400 km) or 0.04 Saturnian radii. This also supports the idea of Uranus' field being developed in its speculated but not yet confirmed ocean of water.

The dipole field of Uranus is affected by the solar wind and gives rise, as at Earth and other planets, to a magnetotail extending away from the Sun. At close to 70 Uranian radii behind the planet the magnetotail was nearly 85

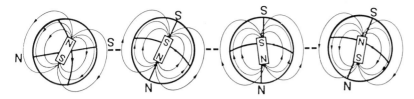

URANUS   SATURN   JUPITER   EARTH

FIGURE 3.3
The offset of the magnetic fields of Earth, Jupiter, Saturn, and Uranus compared with the rotation axis of each planet is shown in this NASA/JPL drawing.

planetary radii in diameter. Its plasma sheet was about 10 radii thick. Contrary to other planets, the magnetotail of Uranus rotates as the planet rotates, as was expected from the planet's unusual orientation in space relative to its orbit. At other times of Uranus' year, when the rotation axis of the planet is not pointing sunward, the tail will not rotate, but will be more akin to the magnetic tails of Earth, Jupiter, and Saturn.

A periodicity in the radio emissions from the Jovian system was attributed to modulation by the satellite Io. Electron flow along the magnetic flux tube linking Io to Jupiter was completed through the ionospheres of Jupiter and Io. Not only did this flow produce bursts of radio energy, but also it might account in part for Io's heating. Flux tubes between Miranda and Uranus along magnetic field lines do not have the potential of inducing Io-type reactions because Miranda has no appreciable atmosphere. Also the interior of Miranda would appear to be very different from that of the volcanic Io. However, there may still be an energy transfer by other processes from the planet's ionosphere to Miranda along the field lines.

One exciting possibility suggested by the large tilt of the Uranian field is that we are witnessing a polarity reversal of the field, an event that occurs every 22 years for the Sun and, as we know from studies of terrestrial rocks, has taken place a number of times during Earth's history. If this were to take place on Earth in our time we would see magnetic compasses pointing to the south instead of to the north. It is a great pity that we have no current plans for another mission to Uranus to investigate the intriguing Uranian magnetosphere at a time 21 years from now, when the rotation axis of the planet will be oriented very differently relative to the Sun.

A most interesting discovery was the presence of electroglow at Uranus. The Dynamic Explorer satellite provided good pictures of terrestrial aurorae and Voyager detected similar auroral regions at Jupiter and Saturn. But at Uranus there was a major surprise. There were expectations that a polar view of Uranus would show a complete polar oval on observations made at a considerable distance from the planet. Such an oval was not seen. At 16 hours before

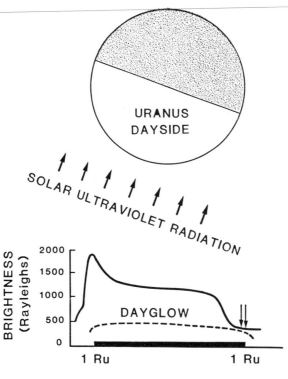

FIGURE 3.4
Hydrogen Lyman alpha emission from Uranus was measured by the spectral scans in ultraviolet across the face of the planet. The ultraviolet electroglow is much greater than the dayglow and completely masks any daytime aurorae. There was a considerable enhancement of the emission from the bright limb of the planet. (NASA/JPL drawing)

closest approach to the planet, a disk scan looked across the face of the planet for ultraviolet emissions. The result was very different from what satellites see of Earth. The emission in hydrogen Lyman alpha was very strong, and this ultraviolet glow was called an electroglow. The phenomenon (See figure 3.4) exhibits strong emission at the limb of the planet (an enhancement of about 2000 Rayleighs) with a lower and constant emission across the face of the planet, then a falling off of emission at the terminator. The limb is referenced to the one bar pressure level of the atmosphere. There could be daylight auroral emissions on Uranus but they appear to be lost in the electroglow and a less intense ultraviolet dayglow.

The electroglow is thought to arise from hydrogen collisions or from electron bombardment of hydrogen molecules above the homopause—the level in the atmosphere above which the composition of the atmosphere begins to change as light gases rise above heavier gases. The electrons have lower energies than those which cause aurorae, but whereas aurorae occur night and day, the electroglow requires sunlight.

Electroglow was first observed on Jupiter but on that planet it is much weaker. It was present on the day sides but not the night sides of Jupiter and Saturn. The conditions seem to be the same at Uranus. Input energy from the Sun (8.5×10^9 watts) does not appear to be sufficient to provide the energy needed to generate the electroglow (1.4×10^{11} watts). The energy source is

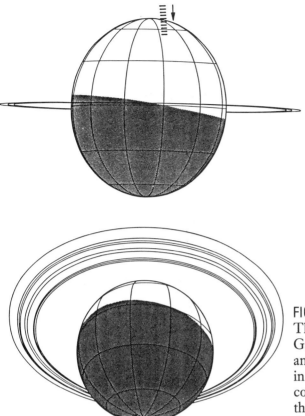

FIGURE 3.5
The occultation path of the star Gamma Persei is shown at ingress and egress. The star was observed in ultraviolet and data were obtained concerning the temperatures over the sunlit pole and the dark pole. In the upper atmosphere the sunlit pole was at a lower temperature than the dark pole.

TO DETERMINE CONSTITUENTS
AND STRUCTURE OF UPPER
ATMOSPHERE

believed to be some kind of coupling to the magnetosphere. Nevertheless an input of solar energy is required for the process to work. Somehow this input energy from the Sun is amplified and reradiated as electroglow. The ultraviolet emissions observed by the International Ultraviolet Explorer satellite are thus now attributed to the electroglow, rather than to aurorae as had at first been speculated.

There is also an ultraviolet dayglow which is obscured by the electroglow. Hydrogen Lyman alpha dayglow from hydrogen atoms is only 30% of the ultraviolet emission from the planet, while the electroglow is 70%. The ultraviolet emissions from hydrogen molecules show an even greater disparity; 1% is dayglow and 99% is electroglow.

Observation, from the spacecraft, of Uranus' occultation of a star (Gamma Persei) (figure 3.5) showed that the upper atmosphere of the planet has a temperature at the sunlit pole of 890° F (477° C) and at the dark pole 1340° F (727° C). It also confirmed the extended hydrogen atmosphere with a homopause, the level below which the atmosphere is mixed and above which gases

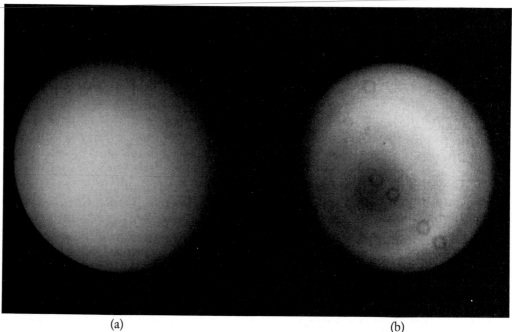

(a)　　　　　　　　　　　　　　　　　　(b)

FIGURE 3.6
(a) As Voyager sped toward Uranus the planet showed an uninteresting image in visible light. (b) Computer processing of the data to enchance the image confirmed the presence of a banded structure and began to show narrow belts closer to the equatorial regions. (Photos. NASA/JPL)

gravitationally separate by masses, located near the level where the pressure is 0.000001 bar.

A nightside aurora was observed on Uranus as was predicted. It is believed to result from precipitation of particles from the magnetotail. The aurora was located at a spot close to the magnetic pole rather than along a pole-circling band. This spot rotated with the planet because of the magnetic pole's offset from the axis of rotation. The nightside aurora was quite distinct and probably results from high-energy electrons penetrating to a great depth into the atmosphere of the planet.

Aurorae were not detected in the visual images of Uranus even though extremely long exposures were tried. Nor were there any of the lightning flashes imaged at Jupiter from the dark side of that planet. Although the ultraviolet instrument could not detect aurorae on the dayside of the planet because they were hidden by the intensity of the electroglow, on the dark side the auroral area extended some 10 degrees around the magnetic pole. Polar aurorae are a minor source of Lyman alpha radiation observed at Earth.

As Voyager sped toward Uranus the planet still showed only a very uninteresting disk in visible light (figure 3.6a). It was only when the images were later

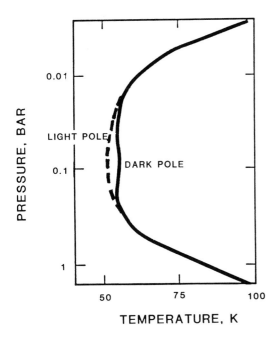

FIGURE 3.7
The temperature as a function of altitude over both poles shows that the stratosphere has a higher temperature over the dark pole. Above and below this region the temperature curves are very similar.

processed by computer and the planet was inspected with other instruments operating beyond the visible spectrum that details of the planet's atmosphere began to materialize (figure 3.6b).

Despite the planet's lack of a large internal heat source and its position at the time of encounter, with one pole pointed toward the Sun, the circulation patterns revealed were very similar to those seen at Jupiter and Saturn—namely, zonal. This implies that the zonal circulation pattern is driven by the rotation of the planet and not by energy received from the Sun. Moreover, temperatures at equator and poles were about the same at levels where atmospheric pressure was between 500 mbar and 1 bar. Somehow, the heat deposited in the atmosphere at the poles is distributed planetwide. Something extracts heat from the polar atmosphere and moves it to lower latitudes. It may be that there is high-altitude movement of air masses from the poles. Complicated and unknown dynamical processes are obviously at work on Uranus. Higher in the atmosphere the temperature falls to −366° F (−221° C), its lowest value, and higher still it rises to 890° F (477° C) in the very high atmosphere. A big surprise was that at pressures just below 1 bar the temperature is higher at the dark pole than at the sunlit pole.

The IRIS experiment provided temperature profiles over both poles as a function of altitude (figure 3.7). Over the dark pole the temperature was higher in the stratosphere than it was over the sunlit pole, but both poles had similar temperatures at higher and lower altitudes, namely, lower and higher pressures. The temperature at various latitudes did not correspond with theoretical

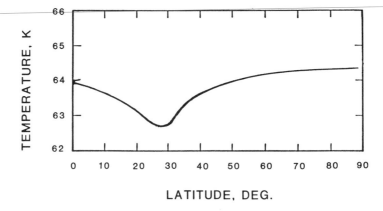

FIGURE 3.8
The equator-to-pole temperature measurements reveal an unexpected feature, a cold region from latitudes 10 to 40 degrees south.

models of an isothermal distribution of heat or a radiative atmosphere with selective absorption of sunlight. Uranus has a cold band between latitudes 10 and 40 degrees south (figure 3.8) which is attributed to dynamic effects of the atmosphere. Temperatures at mid latitudes of Jupiter, Saturn, and Uranus are compared in figure 3.9.

The banded structure of the planet, and individual cloud plumes (figure 3.10), are presumed to be of condensed methane, including methane ice crystals. On Jupiter the visible clouds are ammonia clouds. The narrowness of the Uranian banded structure implies that there is little movement of atmospheric masses toward or away from the poles at cloud level.

In the upper atmosphere there are traces of methane, but it is at higher proportions lower in the atmosphere. Methane gives Uranus its blue-green color, as it absorbs radiation from the red end of the visible spectrum. Another surprising aspect of Uranus was the brightening in methane images toward the limb of the planet caused, it is believed, by the rapid depletion of methane in the atmosphere above the clouds. However, limb brightening toward the equator could result from there being more haze particles in equatorial regions. The methane cloud decks begin at a pressure of about 0.9 bar. Radio science showed that the methane cloud extends to a base at a pressure of about 1.3 bar where the temperature is about −314° F (−192° C). Below the clouds there is probably even more methane in the Uranian atmosphere. There is much more carbon in the atmosphere of Uranus than is found in solar material. This was expected in the outer Solar System.

Radio science showed that the tropopause of Uranus is at a level where atmospheric pressure is about 110 mbars and the temperature is −366° F (−221° C).

The upper atmosphere of Uranus—i.e, the region above the clouds—consists of molecular hydrogen, with some atomic hydrogen, helium, and traces of hydrocarbons. Radio experiments revealed small-scale structures in the stratosphere. Diffusive separation occurs above the homopause which is

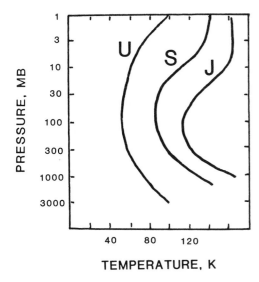

FIGURE 3.9

The temperature of the atmosphere at various pressure levels is compared for the planets Jupiter, Saturn, and Uranus.

located at a level where the pressure is 0.001 mbar. From 1250 to 2175 miles (2000–3500 km) above the appreciable atmosphere (100 mbar level) there is an ionosphere. This ionosphere has two distinct layers of ionized particles at 1250 miles (2000 km) and 2175 miles (3500 km). Such sharply defined ionospheric layers were also observed at Jupiter and Saturn. The Uranian ionosphere probably extends to 6000 miles (10,000 km).

There had been speculation before Voyager's exploration of Uranus that 40% of the Uranian atmosphere might consist of helium, but this was laid to rest when it was found that the hydrogen/helium ratio was approximately that of the Sun. This is also consistent with the possibility that the hydrogen atmosphere of Uranus was derived from the solar nebula and was not an evolved atmosphere from the decomposition of methane early in the evolution of a much hotter planet.

Spectra recorded by the infrared spectrometer and radiometer, coupled with radio science, show clearly that the hydrogen/helium ratio is 0.15 mole fraction (0.26 mass fraction). The abundance of helium is about 10 to 15% of the atmosphere. The helium concentration for Uranus is slightly higher than that for Jupiter (10%) and considerably higher than that for Saturn (2%). It is closer to solar abundance (13%). This could be because helium has differentiated from metallic hydrogen at Saturn and is beginning to differentiate at Jupiter, thereby providing a source of internal heat, whereas this process is not taking place at Uranus and Uranus is not generating internal heat.

The zonal pattern of Uranus changes from very subtle differences between zones within 45 degrees of the pole to many individual and narrow bands closer to the equator. Beyond 35 degrees latitude foreshortening makes it difficult to be certain of the precise nature of the zonal structure in equatorial

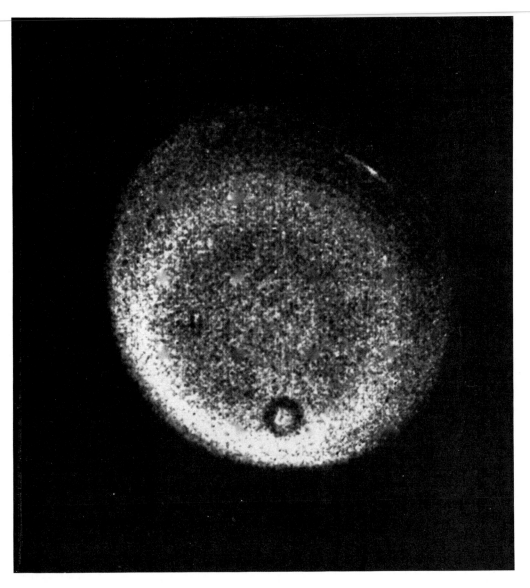

FIGURE 3.10
Processing of the images of Uranus began to reveal a banded structure in more detail including bands and plumes. These latter are assumed to be clouds of methane. The plumes show that winds increase in velocity at higher altitudes. The doughnut-shaped mark near the bottom of the disk is an artifact produced within the imaging system. (Photo. NASA/JPL)

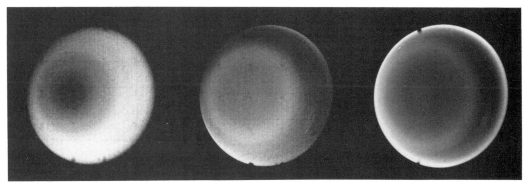

VIOLET   ORANGE   RED

FIGURE 3.11
These three images show Uranus in violet, orange, and red light. Each color reveals different features in the banded structure indicating different substances absorbing the light. The dark polar haze in the left-hand violet picture covers about 20 degrees centered on the pole. (Photo. NASA/JPL)

regions. No evidence of the spot-type weather features of Jupiter and Saturn was seen on Uranus, but a number of small markings lasting a day or so were seen in the images.

All planets with atmospheres have layers of complex molecules that absorb solar ultraviolet in the region from 2500 to 2800 angstroms. Earth and Mars have ozone layers, Venus and Jupiter's volcanic satellite, Io, have layers of sulfur dioxide. The atmospheres of all the giant planets so far explored are now known to have a complex molecular layer of hydrocarbons. On Uranus the layer is at a level where the atmospheric temperature is $-225°$ F $(-143°$ C) and is at a constant altitude from 20 to 85 degrees of latitude. Photochemical actions from incoming solar ultraviolet acting on the molecules would be expected to convert methane to acetylene and polyacetylene. Haze particles exist in the atmosphere above the methane clouds. The haze particles appear to be sized around 3 micrometers diameter.

A dark polar haze cap (figure 3.11) is revealed at violet wavelengths. It covers about 20 degrees centered on the pole and is probably caused by the presence of haze particles that absorb violet light and are concentrated at the pole of rotation by atmospheric circulation rather than by solar heat input. The subsolar point was several degrees from the pole at the time of the observations. There is also a bright ring, concentric with the pole, at 50 degrees latitude which is probably caused by the presence of high methane clouds and haze derived from upwelling air masses. Dark bands at 65 and 20 degrees latitude result from looking deeper into the atmosphere into regions where air masses are descending, haze is less concentrated, and clouds are lower. A cold band is revealed in the temperature data extending from 10 to 40 degrees latitude. In contrast to Jupiter the bright band shows a higher temperature than the dark band. This is probably caused by the difference in latent heats between methane (at Uranus) and ammonia (at Jupiter).

Saturn is more like Uranus than Jupiter in that zones of high albedo (high clouds) do not correspond with regions of high temperature. On Jupiter regions of high albedo and low temperature coincide. Dark regions of Jupiter are thought to be where we can see down into lower, high-temperature clouds.

Bright features at 27 and 35 degrees latitude are thought to be evidence of methane clouds rising several miles above the surrounding clouds (figure 3.12). This is the region of the cold band on the planet detected by the IRIS experiment. The rotation period determined from cloud plumes relative to the rotation period of the planet determined from the magnetic field (17.29 hours) reveals longitudinal winds of 90 to 360 mph (144 to 576 kph). These winds are developed from some process other than heat flow from pole to equator. Wind shears must exist between these regions of differing wind speeds. There is also an increase in wind speed with altitude.

Those plumes observed in the Uranian atmosphere, like those seen in the atmospheres of Jupiter and Saturn, confirm that the winds increase in speed at higher altitudes. One plume was observed over a period of 10 days. It exhibited a core (the rising column of gas) and a streak pulled out from the core by high-level winds. The winds all appear to be in the direction of rotation (see table 3.2). Their prograde direction was a surprise to some scientists.

The dominant circulation pattern in Uranus' atmosphere, like those of Jupiter and Saturn, is predominantly around the planet rather than between poles and equator. Winds generally blow in the direction of rotation (prograde) in contrast with Jupiter and Saturn, where the winds are mainly westerly. Zonal motions are seen also on Earth and Mars, but less pronounced than on the outer planets. To understand these circulations, it is important to note that Uranus receives energy from the Sun at first one and then the other pole, as well as in equatorial regions as with other planets. In spite of this, the circulation is still zonal. It was thus very important that we should have flown by Uranus at a time when the solar input was polar to see this zonal circulation prevailing.

One suggestion is that high-level flows from poles to equator reduce wind speeds at the equator because of the low momentum of the atmospheric masses descending there. One major unknown also is the depth at which the atmosphere in general rotates with the rotation of the inner part of the planet where the magnetic field is generated. Equatorial winds traveling opposite to the planet's rotation would be quite different from Jupiter and Saturn, where the equatorial winds travel in the direction of the planet's rotation.

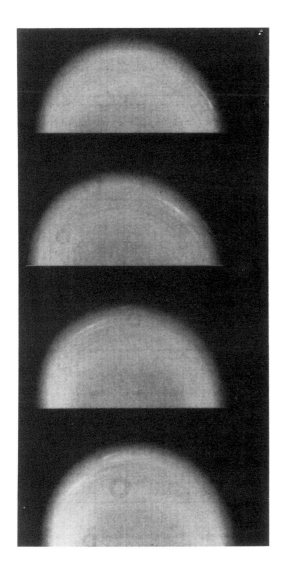

FIGURE 3.12
This series of images shows the motion of clouds around Uranus. These clouds are in the cold band at about 27 degrees and 35 degrees latitude. From their motion wind velocities were measured and found to be prograde; i.e., in the same direction as the direction of rotation of Uranus. (Photo. NASA/JPL)

It may be that the atmosphere rotates more slowly than the magnetic field close to the equator, but faster at higher latitudes. Corotation of the clouds with the magnetic field occurs at about 20 degrees. Maximum differences, and therefore highest wind speeds, occur at about 55 degrees latitude.

Comparisons of the interiors of the giant outer planets are shown in figure 3.13. These planets are thought to have rocky cores but most of their mass consists of ice and gas—the former mainly water and the latter mainly hydrogen. Uranus appears to be a planet with a fairly large proportion of its mass consisting of water or water ice, a very deep ionized ocean within which the circulation necessary for generating the magnetic field takes place. This may account for the displacement of the field because the field is not generated

TABLE 3.2
Uranus Winds

Latitude	Rotation Period hours	Calculated Speed	
		mph	kph
26.7	16.9	30	50
34.8	16.3	215	345
40.2	16.0	298	478

within the rocky core. The atmospheric studies have shown that the percentage of helium is close to that of solar material. The overall density of the planet implies a greater proportion of denser inner materials than Saturn and Jupiter. While the rocky cores may all be about the same size, Uranus has a much deeper water "ocean" compared with the overall size of the planet than do Jupiter and Saturn, and a smaller hydrogen-rich atmospheric shell. The rocky cores and icy oceans of Jupiter, Saturn, Uranus, and Neptune may all be about the same mass, but Jupiter and Saturn have gathered much more light gases than has Uranus, and presumably Neptune.

This is because in the outer regions of the Solar System when Uranus formed, rocky materials such as silicates and iron, which would have condensed at the temperature in the solar nebula, were readily incorporated into the structure of the planet while gases such as hydrogen and helium were not so easily captured. Ices such as water and methane were captured more easily than the gases. Thus Uranus would be expected to have a rocky core, a deep icy ocean, and a relatively small hydrogen atmosphere. With this assumption the ratio of masses of ice and rock can be calculated. The planet probably has 43% of its mass as a rocky core, 38% as icy ocean, and about 19% as the gaseous atmosphere.

A major mystery about Uranus is the low heat flux from the interior; less than 180 erg/cm^2/s. The Earth has a heat flux from its interior of 62 erg/cm^2/s, Jupiter a flux of 5600, and Saturn one of 2000. Neptune appears to have a flux of 285. Jupiter derives its heat flow from the cooling of its interor, Saturn from cooling and from differentiation of helium. It is likely that the internal temperature of Uranus is several thousand degrees, Jupiter's is 30,000 degrees, and Saturn's is somewhat less than Jupiter's. Heat stored in the central core at the formation of the planet would not be able to escape because of the low conductivity of the icy mantle. A temperature of 6000–8000° K could exist in the core, while at the bottom of the hydrogen-rich atmosphere where it interfaces with the icy ocean, compression temperatures of 2000° to 3000° K might be present. These temperatures, with the icy mantle between 3000° and 6000°K, would mean that the icy mantle would be a liquid ocean, at

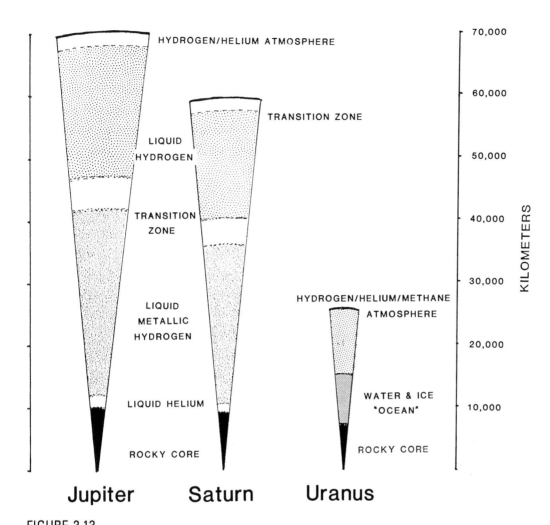

Jupiter Saturn Uranus

FIGURE 3.13
Comparisons of the interiors of the big planets according to a present understanding
of the planets Jupiter, Saturn and Uranus. Some of the Voyager data, however, suggest
that the water and ice "ocean" of Uranus mixes with atmosphere and there is no clear
boundary as shown on this drawing.

least in part. This is important because it would allow for the dynamo action
needed to create the observed magnetic field.

Also these temperatures are sufficient to account for heat flow from the
interior of the planet over millennia since the planet was formed. No heat flow
appears from Uranus, some is observed at Neptune. This is because the temperature of Uranus produced by incoming solar radiation is higher than that at
Neptune, which is more distant from the Sun. Both planets are probably at
similar internal temperatures but the temperature difference between interior
and exterior is less at Uranus than at Neptune.

Deep within the large outer planets the inner regions rotate like a solid body.

This rotation period is the period of rotation of the planet's magnetic field. It is the basic rotation rate of the planet as compared with the different rotation rates of the observed clouds and the zonal wind systems which are sometimes greater than or less than the true period of rotation. An extreme case is Venus whose atmospheric markings rotate in 4½ days compared with the 243-day rotation period of the planet itself. In the case of Venus, however, the driving energy for the atmospheric circulation is derived from the Sun, rather than originating within the planet.

Measurements of the trajectory of Voyager past Uranus allowed the shape and gravity field of the planet to be determined more precisely. The imaging science places the equatorial radius of Uranus as about 15,940 miles (25,650 km). The trajectory of the flyby and the illumination of only the southern hemisphere of the planet were not suitable for obtaining a good measurement of polar flattening. Analysis of the gravity field has allowed an updated view of the planet's interior. The three-layer, spherically symmetric structure appears to hold. The inner core of several Earth masses consists of magnesium silicates and iron. Surrounding this is the mantle of water and ices of water and other constituents such as methane and ammonia. The outer layer is the hydrogen-rich atmosphere. The water "ocean" provides the electrically conductive fluid mass to develop the intrinsic magnetic field of Uranus. High-pressure water provides sufficient conductivity for a dynamo. The measured field and its orientation lead to the conclusion that the lower region of the ocean is the most likely place where the dynamo is located. However, as mentioned previously, the peculiar orientation of the field may be because we are measuring it in the process of a field reversal of the type Earth has experienced many times during its history. If so, circulation in a liquid iron inner core, similar to Earth, could be providing the dynamo for the magnetic field.

No light has been thrown as yet upon why Uranus rotates on its side, although the consensus of speculation is that there was a catastrophic event after the planet was formed. Similarly, thoughts on the origin of the Earth's Moon have been swinging more toward a catastrophic scenario; there has been speculation that a body the size of Mars struck the Earth and created the Moon. It appears realistic to suppose that in the early days of the Solar System there were many catastrophic impacts. Even today the possibility exists of Earth-orbit-crossing asteroids colliding with the Earth with catastrophic consequences greater than any man-made events from thermonuclear exchanges. Ring systems of the outer planets may be evidence of such catastrophic events, as is discussed in the next chapter. The rings of Saturn, Jupiter, and Uranus may be relatively short-lived appendages to those planets, and may not have been in existence for more than a few hundred million years. This is because forces are at work disrupting planetary rings—the drag of extended planetary atmospheres and gravitational effects. The extended hydrogen coma surrounding Uranus, for example, extracts kinetic energy from the ring particles so that they spiral down toward the planet. This removes small particles first. Gravitational interactions also remove particles by collisions.

While many questions have been answered about Uranus as a planet, other more detailed questions are being raised. Undoubtedly a clearer picture will emerge when Voyager flies past Neptune and the two outer giants can be compared in greater detail.

4

RINGS AND SATELLITES

Algol was the *lucida* of the Gorgon Medusa. To gaze upon it was to be turned to stone. Algol was also the name given by Ptolemy to the star known as Beta Persei, a variable star in the constellation of Perseus. This Gorgon's light was used by Voyager to determine the size of the "stones" in the rings of Uranus, which turned out to be big, coal-black boulders.

On January 24, 1986, the photopolarimeter recorded the light from Beta Persei as the motion of the spacecraft caused the star to move behind the rings. Other occultations were also observed; of Sigma Sagittarius for the rings, and Gamma Pegasi for the planet's atmosphere. Figure 4.1 shows the path of Beta Persei relative to the planet and its rings as seen from the spacecraft; figure 4.2 shows the path of Sigma Sagittarius. This latter occultation, a 30-minute grazing occultation along an arc of the rings, took place on January 23–24, when Voyager was just over 20 planetary radii from Uranus during the approach to the planet and it provided details of the epsilon and delta rings. The Beta Persei occultation on January 24 took place as the spacecraft sped away from the planet when Voyager was just over 10 planetary radii beyond its closest approach. The occultation occurred more quickly; taking just under fifteen minutes. But it covered all the rings on either side of the planet and consisted of an ingress sequence and an egress sequence almost on opposite parts of the ring system. The paths are compared in figure 4.3, which shows a plan view of the Uranian system.

The intensity of light from the star was sampled 100 times a second as it passed through the rings to the photopolarimeter. The occultations provided much information about the orientation, the structure, and the components of

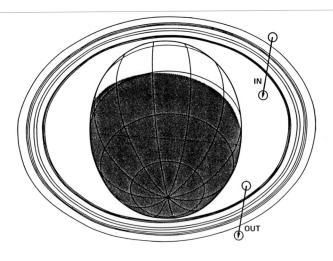

FIGURE 4.1
The path of Beta Persei relative to Uranus and its ring system as observed from the Voyager spacecraft. (NASA)

the rings. A multitude of detailed structure was revealed that cannot be seen in the broad features determined from occultation observations based on Earth. Four profiles were obtained of the epsilon and delta rings, and two of all the other rings. Examples for several rings are shown in figures 4.4 and 4.5.

Several thin ring fragments were found and an additional ring was dis-

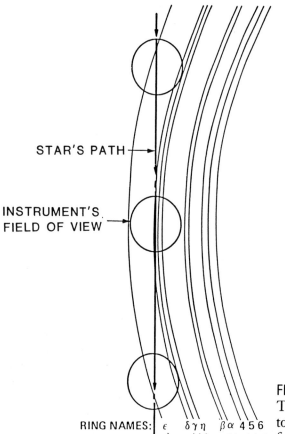

STAR'S PATH

INSTRUMENT'S FIELD OF VIEW

RING NAMES: ϵ $\delta\gamma\eta$ $\beta\alpha$ 4 5 6

FIGURE 4.2
The path of Sigma Sagittarius relative to the rings of Uranus as observed from the Voyager spacecraft. (NASA)

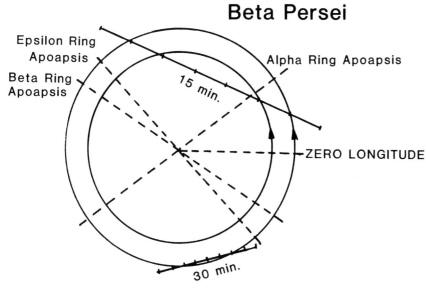

Beta Persei

Epsilon Ring
Apoapsis

Alpha Ring Apoapsis

Beta Ring
Apoapsis

15 min.

ZERO LONGITUDE

30 min.

Sigma Sagittarius

FIGURE 4.3
Comparison of the occultation paths as seen in a plan projection of the ring system. The Beta Persei occultation passed completely through the ring system twice, while the Sigma Sagittarius sampled only the outer part of the ring system. (NASA)

covered. Strangely, however, this ring does not appear in any of the occultation data gathered at Earth. It may be a transient ring. Such rings could be formed by the impact of a large meteoroid on a small satellite orbiting within the Roche limit. Satellites within that distance from a planet—2.44 times the planet's radius—tend toward breakup rather than accretion if hit by an asteroid-type

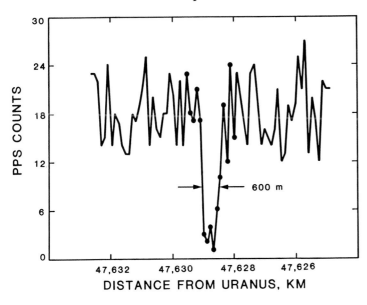

FIGURE 4.4
The variation in starlight as Beta Persei ingressed the rings shows details of the Gamma Ring and a width of only 600 meters. (NASA)

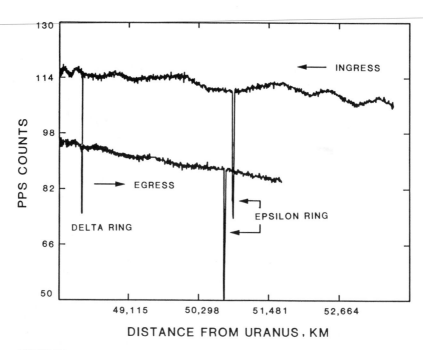

FIGURE 4.5
The Sigma Sagittarius ingress and egress results for the Epsilon Ring show a difference in radial distance from the planet even at the short arc of rings sampled. The Delta Ring, by contrast, does not show eccentricity over this arc. (NASA)

body or by meteoroids. The ring systems discovered so far—those of Jupiter, Saturn, and Uranus—are all within the Roche limits for their primaries

The total mass of all the Uranian rings is very small compared with the ring system of Saturn. The complete Uranian system probably has less mass than the ring materials in the low-density "gaps" in Saturn's rings, but the Uranian ring system is complex, nevertheless. The occultation data revealed the presence of many faint ringlets around the planet. Also there is a very faint ring of dust extending downward, as at Jupiter and Saturn, toward the top of the planet's atmosphere. This was recorded on a long-exposure image obtained close to the plane of the ring system (figure 4.6).

In addition to the optical occultation data, imaging data and radio data (the latter derived from radio signals from Voyager passing through the rings on their way to Earth) provided important information about the ring system. While optical occultation observations provide information about small ring particles, radio occultation observations provide information about larger particles. Imaging provided fundamental data on ring structure and on the sizes of ring particles.

When Voyager crossed the plane of the rings, the plasma wave experiment recorded an intense burst of activity, presumably from dust particles, which

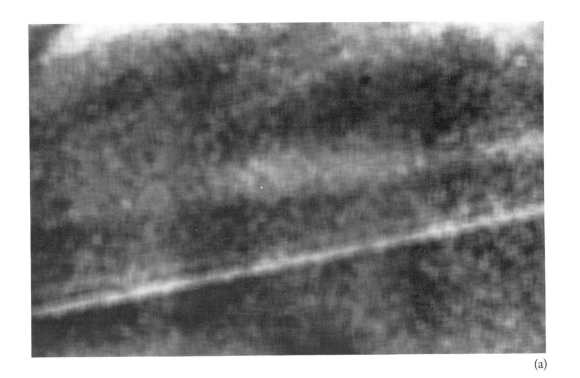

(a)

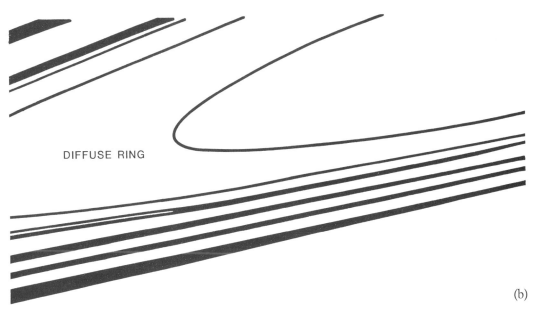

DIFFUSE RING

(b)

FIGURE 4.6
(a) A long exposure image of the rings made close to the ring plane crossing revealed material within the known ring system. This material is extremely diffuse and it extends toward the planet. (b) A line drawing of the same area shows the location of the diffuse wide ring. (Photo. NASA/JPL)

TABLE 4.1
The Rings of Uranus

Name	Mean Diameter		Approx. Width		Notes
	miles	km	miles	km	
Low Density	>32,000	>51,000	1800	3000	Very diffuse
Epsilon	31,808	51,188	12–60	20–95	2 components
New Ring	31,480	50,660	10	16	
1986U1R	31,095	50,040	24	38	Possibly arcs
Dark lane	30,884	49,700	—	—	Satellite swept
Delta	30,020	48,310	4	7	3 components
Gamma	29,616	47,660	1	2	
Eta	29,330	47,200	2	3	2 components
New Ring	28,420	45,736	1	2	Very faint
Beta	28,398	45,700	3–8	5–12	3 components
Alpha	27,813	44,758	3–8	5–12	4 components
4	26,478	42,610	2	3	
5	26,270	42,275	2	3	
6	26,022	41,877	2	3	
1986U2R	<24,730	<39,800	1500	2500	Very diffuse
New Ring	23,787	38,280	1	1	

Except for the eta ring, all are eccentric, and most are slightly inclined to the equatorial plane of Uranus. Several incomplete rings, ring arcs, were discovered. (Based on optical imaging, photopolarimeter, and radio data.)

lasted some six minutes. This indicated the presence of ring material far beyond the visible rings revealed by imaging and occultation observations. A similar increase in plasma wave activity had been recorded earlier in the crossing of the ring plane of Saturn.

Details of the ring system are given in table 4.1. The brightest ring, epsilon, is now known to consist of at least two rings, plus some fine-scale structure almost at the limits of resolution. The outer edge of the ring is very sharply defined, the inner edge less so; transition from ring to "empty" space being only a matter of 165 feet (50 meters) and 1640 feet (500 meters) respectively. The radio data indicate that the ring particles are 3 feet (1 meter) or more in diameter since signals at both frequencies passed through the rings in the same way.

On January 24, 1986, as the spacecraft moved away on the far side of Uranus from Earth, radio occultation data about the rings were obtained on opposite sides of the planet as the occultation trace first entered the rings at one side, then passed behind the planet, finally passed behind the rings again on the far side, see figure 4.7.

The radio waves from Voyager were at 3.6 cm and 13 cm. The data indicate that the ring system contains much material of a size comparable with the wavelength of the radio waves—i.e., 3 to 13 cm, or larger. A new scattering

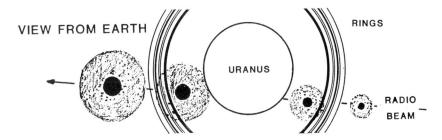

VIEW FROM EARTH

RINGS

URANUS

RADIO BEAM

FIGURE 4.7
Elements of the radio occultation as viewed from Earth. The spacecraft was speeding away behind Uranus during the occultation, so the apparent diameter of the radio beam relative to the ring system was increasing. The dark spot is the beam of the high frequency, the diffuse spot that of the lower frequency. (NASA)

phenomenon was also observed, unlike anything experienced at Jupiter or Saturn. This occurred at 3.6 cm in connection with the epsilon ring, and it is thought to be caused by a scattering center, a group of bodies rather than a single large body immersed in the rings, or some structure oriented radially within the ring.

The radio data also suggest that the epsilon ring in which there appeared much structure (figure 4.8), consists of layers of particles rather than a single layer.

At one occultation the delta ring exhibited three major components. Strangely, at another occultation, at a different part of the ring, only one component was evident. The next ring inward, the alpha ring, has three faint broad components as well as a bright narrow one. The beta ring has two broad components—5 and 7 miles (8 and 11 km) wide. The gamma ring is very narrow and has sharp edges. The width varies around the planet. Rings 4, 5

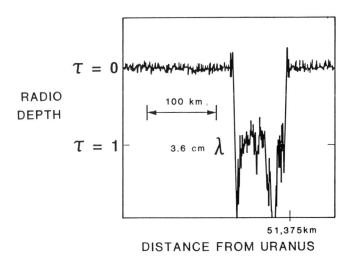

RADIO DEPTH

$\tau = 0$

$\tau = 1$

100 km

3.6 cm λ

51,375km

DISTANCE FROM URANUS

FIGURE 4.8
This plot shows the details of the Epsilon Ring obtained by the radio occultation experiment at the time of egress. Much structure is shown within the ring at the 3.6 cm wavelength. The width of the ring is just less than 100 km. (NASA)

RINGS AND SATELLITES

107

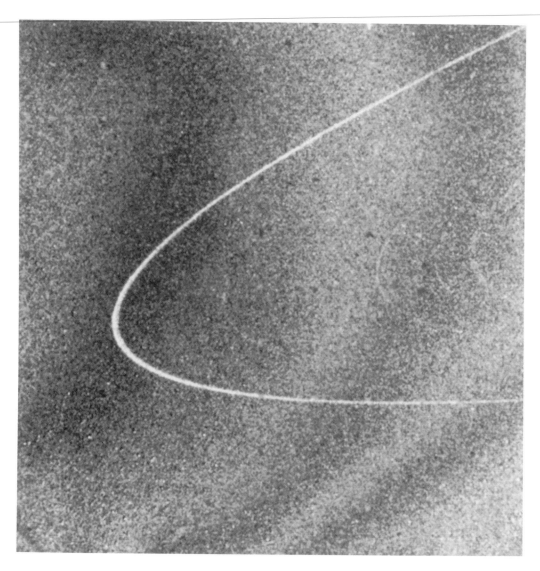

FIGURE 4.9
This image of the outer part of the ring system obtained January 24, 1986 just 11 minutes after passing through the ring plane. Between the bright outermost Epsilon Ring and the Delta Ring is a very faint, newly discovered ring which was named 1986U1R and is marked by the arrow. (Photo. NASA/JPL)

and 6 are also very narrow, as is the newly discovered 1986U1R ring between the epsilon and the delta rings (figure 4.9).

With the exception of the eta ring, all the rings are eccentric. The eccentricities are illustrated in figure 4.10. The epsilon ring and ring 5 are the most eccentric. Rings eta, gamma, and epsilon are close to the plane of the planet's equator; the other rings are all inclined, with 5, 6, and 7 having the greatest inclinations, but all less than .1 degrees.

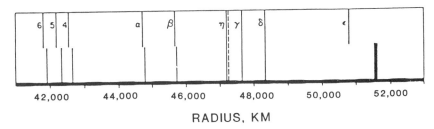

FIGURE 4.10
All the rings with their ellipticities are shown in this diagram. (NASA)

The rings differ from each other greatly in brightness, structure, and width (figure 4.11), although all are narrow and widely spaced from each other when compared with the many rings of Saturn. Inside and outside the main rings there appear to be wide rings of small particles reaching out toward Miranda's orbit and down toward the top of the atmosphere. These rings were not previously suspected in Earth-based observations. The material in them is extremely diffuse, but firm evidence for the presence of these extended diffuse rings was obtained by the imaging system close to the ring plane (the inner diffuse ring) and by the plasma wave instrument when the spacecraft crossed the ring plane (the outer diffuse ring). A diffuse dust cloud resulting from collisions of ring particles may be present throughout the ring plane and extend possibly several thousand kilometers above and below it. Generally, however, the rings are comparatively free of dust and appear to consist of swarms of large boulders with diameters greater than 3 feet (1 meter).

All the rings exhibit some kind of structure, the epsilon ring (figure 4.12) has two distinct features with a less intense ring between, the eta ring (figure 4.13) has a diffuse ring some 36 miles (60 km) wide, as well as the more clearly defined narrow ring; and delta has three major components. There are variations in width and optical depth around each ring. The epsilon ring varies from a 12 mile (20 km) width at its closest part to the planet (periapse) to about 60 miles (95 km) at the most distant part of the ring from the planet. Periapse and apoapse differ by approximately 275 miles (440 km). The alpha and beta ring thicknesses both vary from 3 to 7 miles (5 to 12 km), but the smallest and largest thicknesses do not coincide with the periapses and apoapses but are displaced some 30 degrees from those points.

It seems quite clear from the Voyager data that the rings are not symmetrical, but brighter and thicker at some parts than at others. The material in the rings is not spread evenly around them. The eta ring, for example, has both a narrow and a broad component, but the narrow component, which has been detected from Earth, was not in evidence at one of the occultations observed by the spacecraft; the ingress of Beta Persei.

Arc rings have been suggested as an explanation for some of the peculiarities

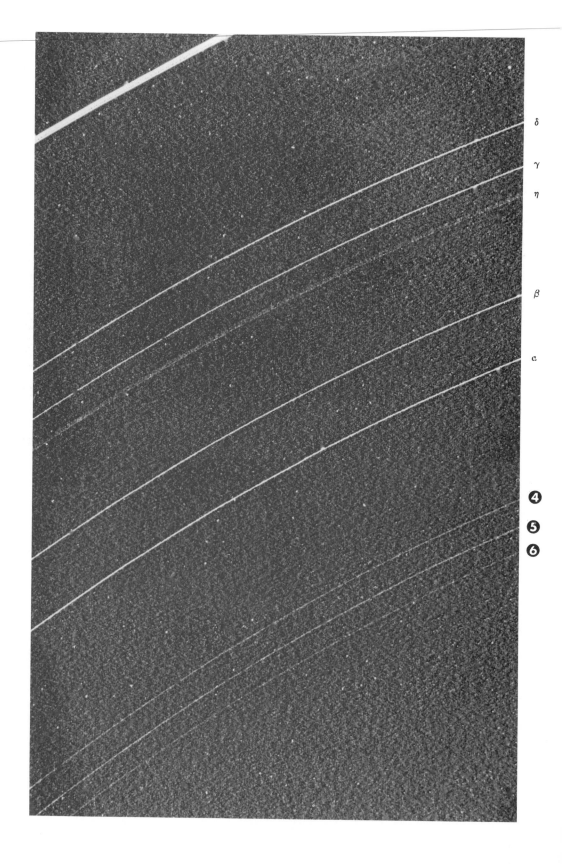

δ
γ
η

β

ε

❹
❺
❻

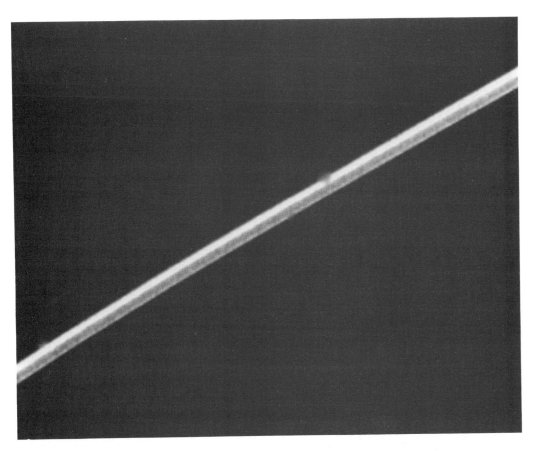

FIGURE 4.12
High-resolution image of the bright Epsilon Ring obtained on January 23, 1986 from a distance of 690,000 miles (1.12 million km). The ring is about 60 miles (100 km) wide in this view and clearly shows a structural variation. There is a bright, broad outer component about 25 miles (40 km) wide, a darker middle region of about the same width, and a bright inner narrow region about 9 miles (15 km) wide. The two fuzzy splotches on the outer part of the ring were generated during processing of the image. (Photo. NASA/JPL)

seen in Earth-based occultation data on Neptune. These arcs were detected at Uranus not only within the ring system but also beyond the epsilon ring.

Before the encounter there was much speculation that shepherding satellites might have been flanking the rings and herding the particles into the narrow rings detected at Uranus. However, despite searches by the imaging team, only

FIGURE 4.11
All the rings are shown in this image obtained January 23, 1986 and consisting of a two-frame mosaic. The rings are identified along the margin of the picture. (Photo. NASA/JPL)

FIGURE 4.13
Taken at a distance of 690,000 miles (1.12 million km) this image shows the broad outer component and the narrow inner component of the Eta Ring. The broad component is considerably more transparent than the dense, narrow inner component. (Photo. NASA/JPL)

two shepherding satellites were found, 1986U7 at 30,635 miles (49,300 km), and 1986U8 at 33,120 miles (53,300 km) on either side of the bright epsilon ring (figure 4.14). Satellites would tend to sweep particles and leave a gap in the ring system. This is apparent for 1986U7, and some of the photopolarimeter data suggest other gaps may be present in which satellites too small to be

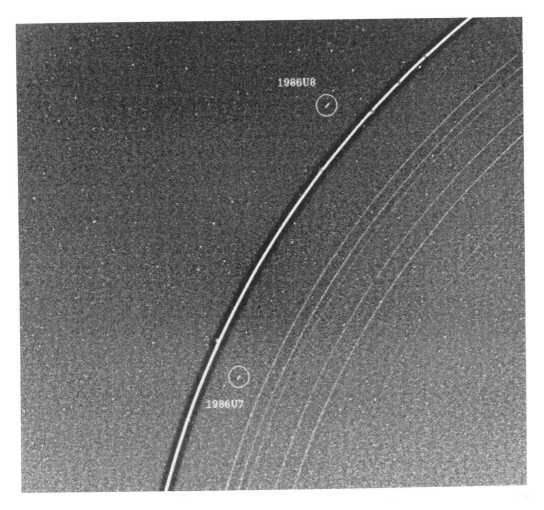

FIGURE 4.14
The Voyager imaging team discovered two satellites acting as shepherds to the Epsilon Ring. This image, obtained on January 21, 1986, shows these satellites and all the nine rings. The discovery of these shepherding satellites was a major advance in understanding of ring systems and how narrow rings are kept from spreading out. At Saturn, for example, the narrow F Ring also has two shepherds. A disappointment at Uranus was, however, that no other shepherds were found for the other narrow rings of that planet. (Photo. NASA/JPL)

detected visually may be sweeping up ring material. The presence of these satellites, however, is difficult to confirm.

The effects of shepherding to confine a ring can theoretically also be achieved by a swarm of small satellites as well as by a single satellite. This raises the possibility of shepherding swarms at Uranus that cannot be seen on the images returned from the planet.

Unfortunately, even in the best images of the rings obtained by Voyager, most of the rings could not be resolved into greater detail. But some variations in brightness along the rings are evident. By imaging the rings in different colors (green, violet, and neutral) the scientists of the imaging team were able to show that the epsilon ring is grey, and there were faint indications of differing shades for other rings. The problem is that the light reflected from the rings is so extremely faint that it proves very difficult to identify color differences as was done at Saturn.

An important feature of the Uranian system is the different color of the particles compared with the ring particles of Jupiter and Saturn. The latter are bright, but the Uranian rings are composed of extremely dull particles. The particles of Jupiter's and Saturn's rings tend toward a reddish color. The Uranian particles are very dark grey. The Jovian and Saturnian particles are thought to be water ice; perhaps the Uranian particles are methane ice which has been blackened by exposure to energetic particles of the radiation belts.

An image taken at a high phase angle (172.5 degrees) with an extremely long exposure (96 seconds) shows the rings backlighted (figure 4.15). The spacecraft was in the shadow of the planet at the time the image was obtained, so conditions were ideal for revealing fine faint detail. At first the image seemed to bear no relationship to those taken by reflected light from the rings, but by calculating where the epsilon ring and other bright rings should appear and then matching these positions with the rings on the image, the other ten rings were connected with features on the backlighted image. While the ten known rings were subdued, a number of other rings appeared bright under the backlighting conditions of observation. These are presumably rings of dust particles which would forward scatter sunlight, whereas the big boulders of the previously known rings do not forward scatter very well. The backlighted rings, only visible by forward scattered sunlight, are similar to the rings of Jupiter and some of the rings of Saturn.

The dust rings of Uranus appear from analysis of the imaging data to be somewhat analogous to the dusty rings of Saturn. They have multiple fine bands like the D ring of Saturn which is a very tenuous ring between the innermost bright ring visible from Earth and the cloud tops of the planet itself. The D ring actually consists of a number of ringlets, as do the Uranian dust rings.

Uranus has probably captured many long-period comets which have become short-period, with their apoapses about the distance of Uranus from the Sun. These cometary bodies must periodically impact the satellites of Uranus and spread debris. Moreover, large ring particles are colliding with each other. It has been estimated that small satellites within the orbit of Miranda may have lifetimes measured in less than a few billion years, perhaps less than one billion. In fact the small satellites may have originated from the disruption of a larger satellite. Satellites can also accrete from smaller particles, and this pro-

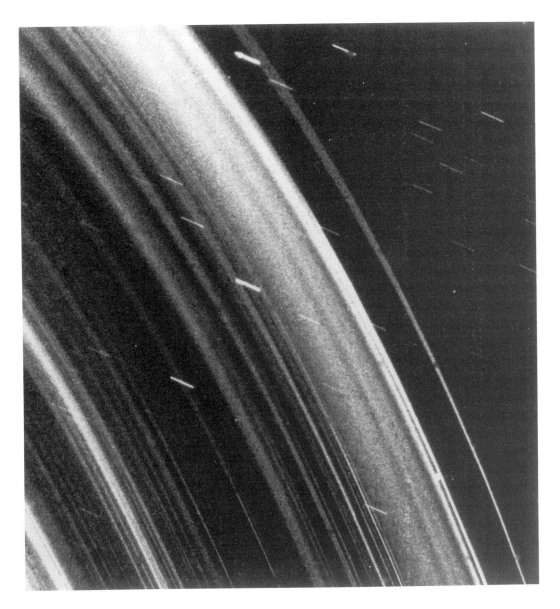

FIGURE 4.15

This dramatic image reveals an almost continuous distribution of small particles throughout the Uranian ring system. It was obtained while Voyager was in the shadow of Uranus at a distance of 147,000 miles (236,000 km) at a high phase angle on the far side of the planet from Sun and Earth. The rings were backlighted, which provided an image of lanes of fine dust not visible at other lighting angles. An extremely long exposure of 96 seconds was needed to obtain this image, which gives rise to the trails of the background stars. Some of the brightest features of this image are dust lanes, which cannot be seen under front lighting. (Photo. NASA/JPL)

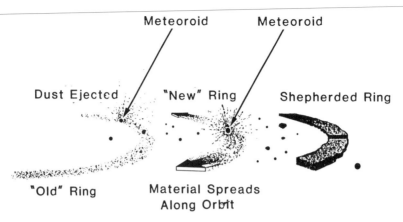

Meteoroid   Meteoroid

Dust Ejected   "New" Ring   Shepherded Ring

"Old" Ring   Material Spreads Along Orbit

FIGURE 4.16

A possible mechanism for creation of rings is shown in this NASA drawing. The epsilon ring is maintained by the shepherding satellites. Other more transient rings originate as meteoroids encounter small satellites and break them up. The material spreads into an orbit first as a ring segment and then into a complete ring. Further collisions produce dust which is gradually removed from the system by atmospheric drag.

cess has probably been at work in the Uranian system, even for some of the larger satellites which seem to show surfaces indicating that the body has been assembled from smaller bodies that accreted together.

A surprise about Uranus was the absence of other small satellites, which were at one time postulated as needed to keep the rings in place. The imaging experimenters state that objects as small as 9.5 miles (15 km) in diameter would have been visible if they had been present unless they were even darker than those satellites discovered.

Even though only two shepherding satellites were found in the Uranian system, the rings must have some mechanism to confine them into their orbits. Otherwise they would dissipate rapidly because of energy losses due to collisions, and aerodynamic drag of the outer hydrogen corona. It is not known what this mechanism is. Without it, however, the ring system of Uranus would have to be regarded as a very transient phenomenon on an astronomical time scale; possibly it is a young system created by a collision between an asteroid or a comet and a satellite. If so, the Uranian ring system will gradually disappear.

Ring particles may last for only a few million years before being broken apart by collisions. Thus larger satellites break up into smaller ones, which in turn break up into particles to form rings (figure 4.16). The ring particles gradually break down into smaller ones, and when these reach dust size they are removed from the ring system by coronal drag and other forces. The Uranian rings are believed to be relatively young because they are inhomogeneous and lack many

TABLE 4.2
Newly Discovered Small Satellites of Uranus

Name	Diameter		Orbit Radius		Orbit Period	
	miles	kilometers	miles	kilometers		
1985U1	100	160	53,430	85,980	18h	17m
1986U5	30	50	46,670	75,100	14	56
1986U4	30	50	43,450	69,920	13	24
1986U1	56	90	41,070	66,090	12	19
1986U2	43	70	40,000	64,350	11	50
1986U6	30	50	38,960	62,700	11	24
1986U3	43	70	38,370	61,750	11	06
1986U9	15	25	36,790	59,200	10	00
1986U8	12	20	33,120	53,300	8	55
1986U7	9	15	30,880	49,700	7	55

small particles. The rings of Uranus may thus be very variable features (as may those of Saturn and Jupiter) when viewed over time spans of millions of years.

The Voyager encounter led to the discovery of ten previously unknown satellites listed in table 4.2. These satellites were discovered by the imaging team by close inspection of the many images returned from Uranus (figure 4.17). These newly discovered satellites all orbit within the orbit of Miranda, which leads to speculations that they originated from a larger satellite which was broken apart by collision with a large body entering the Uranian system. The impact of a body on a satellite is expected to break up that satellite if the energy is sufficient to form a crater whose diameter is comparable to that of the satellite itself. Thus, the small inner satellites of Uranus and the ring system may have originated from the breakup of a larger satellite and subsequent collisions between the particles resulting from the disruptive collision.

The largest of these small satellites is 1985U1, which orbits at 53,430 miles (86,000 km). Its diameter is 100 miles (160 km). The other nine are all much smaller and orbit between it and the epsilon ring, with the exception of 1986U7, (figure 4.18) which is the innermost of the group and appears to be the inner shepherding satellite of the epsilon ring. Its orbit is at 30,880 miles (49,700 km).

These newly discovered satellites and the previously known five larger satellites are all in orbits that are approximately circular and in the same plane. Inspection of the images of the five large satellites showed that they present one face to Uranus. This was expected.

The masses of the large satellites, which had been very difficult to determine with any degree of precision from Earth, were refined. This is important

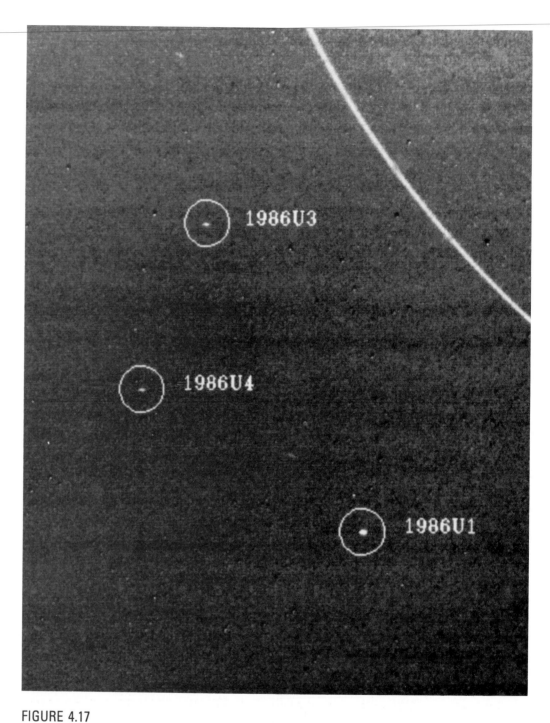

FIGURE 4.17
Close inspection of the many images returned by Voyager led to the discovery of more small satellites. Here are three of them orbiting within the orbit of Miranda but outside the ring system. (Photo. NASA/JPL)

FIGURE 4.18
The innermost of the newly discovered satellites is one of the shepherds, 1986U7, shown here elongated because of the exposure time. On this image, too, the bright Epsilon Ring is resolved into its bright and dark components. (Photo. NASA/JPL)

because masses, coupled with the refinements in measurements of satellites' diameters, produced a better estimate of the densities. Densities are needed to understand their internal structure and composition. The mass of Miranda was determined from its gravitational pull distorting the path of the spacecraft as it made its very close approach to the tiny satellite. The masses of the other satellites were not so easily obtained because the spacecraft did not approach very close to them and also passed almost vertically through the plane of their orbits. However, as Voyager sped toward Uranus a number of images were obtained to determine the precise position of the satellites relative to the background of stars. These optical navigation observations enabled celestial dynamic specialists to compute the orbits of the satellites more accurately, from

TABLE 4.3

Physical Characteristics of Major Satellites
(As accepted pre and post Voyager's encounter)

Name	Diameter miles PRE-V	Diameter miles POST-V	Mass Moon/100 PRE-V	Mass Moon/100 POST-V	Density gm/cm³ PRE-V	Density gm/cm³ POST-V
Miranda	310	300	.23	.18	3.0	1.3
Ariel	825	720	2.13	3.32	1.3	1.7
Umbriel	690	740	1.36	3.14	1.4	1.4
Titania	995	1000	8.05	8.56	2.7	1.6
Oberon	1010	963	8.19	7.19	2.6	1.5

Based on radio tracking and optical navigation results.

which their mutual disturbances could be derived. In turn, this allowed masses to be determined. Radio tracking was also used to supplement the optical data. These masses are shown in table 4.3, which compares them with the masses derived from earlier Earth-based observations.

All the small satellites have albedos close to that of the ring particles; they have extremely dark surfaces. This is in sharp contrast to the small satellites of Saturn, which are quite bright objects. One problem in interpreting what was seen of the Uranian satellites is that they were all presenting one hemisphere to Voyager's cameras. Elsewhere in the Solar System we have often been surprised by great differences between different hemispheres of a single planet. Earth, for example, has continental masses dominating one hemisphere and the Pacific Ocean dominating another. Earth's polar regions are also quite different from each other, one an ocean, the other a continent. The Moon has a hemisphere of great plains and one of cratered highlands. Mars has the great volcanic region of Tharsis, entirely different from another hemisphere of ancient cratered terrain. While there were opportunities to see the hemispheres of other planets and of the satellites of Jupiter and Saturn, there can be no opportunity to look at the other hemispheres of the Uranian satellites until Uranus' other pole begins to point toward the Sun, some 40 years from now.

One of the ways to investigate the physical nature of the surfaces of satellites is to study the degree of polarization of the sunlight reflected from them, as was done extensively in studies of Earth's Moon before spacecraft enabled us to obtain photographs on the surface and astronauts obtained samples of the surface materials.

When sunlight hits a planetary surface, some is absorbed, some is reflected, and some is scattered in many directions. The interaction of sunlight with the satellite's surface causes the light coming from the surface to have different intensities in different planes of polarization. The proportion of polarized light to the total amount of light provides information about the polarization characteristics of the surface materials which can be compared with the polarization of materials in a laboratory.

In 1934 B. Lyot founded the basics of this type of work when he commenced in-depth studies of the materials of the lunar surface, and discovered characteristics of a granular, opaque surface material of the lunar regolith somewhat similar to volcanic ash. This was later shown by lunar landings to be a regolith of fine material generated by multiple impacts of meteorites.

Another important characteristic of the Moon's surface was that it was many times brighter at full moon than at first and last quarters, although the visible surface illuminated at full moon was only twice that at the quarters. The shielding of parts of the surface by shadows was not sufficient if the effect was attributed to shadows from large objects. But the full moon brightening—sharp backscattering like a beaded movie screen and often referred to as an opposition brightness surge—might be accounted for by a complex microstructure, with many deep pits and microscopic but steep hills casting shadows at other than full phase, and an almost spongy-type of material. Many asteroids also exhibited the opposition brightening effect, which was attributed to their surfaces being covered with a thin layer of finely powdered material.

Observers of the large Uranian satellites from Earth had seen an opposition brightness surge which, especially at the near infrared part of the spectrum, was unusually high compared with those seen on other bodies of the Solar System—even larger than the opposition brightness surge of Saturn's rings, which had been the highest observed previously. This implied a surface of fine particles on all the satellites, even though they were known to have icy surfaces.

The Voyager results confirmed the Earth-based observations of an opposition brightness surge which indicated that the surfaces are less compacted than the lunar regolith—certainly for Titania, and presumably for the other satellites as well. The phase angle observations also showed that on a larger scale the surface of Titania is about as rough as the surface of Earth's Moon. This was born out by the images of Titania's surface, which revealed abundant impact craters pockmarking the surface, much like the ancient cratered surface of the lunar highlands. While high-phase-angle data were obtained for Titania to provide the comparison, such data were not obtainable for the other satellites because Voyager's path through the Uranian system did not provide a suitable geometry.

What is the surface material? All the Uranian satellites are darker than the satellites of Saturn except for Phoebe and the dark hemisphere of Iapetus. Also the color of the surfaces is grey, whereas many other Solar System satellites have a tendency toward redness. Of all the Uranian satellites only Oberon has a certain amount of redness that tends to resemble the color of Phoebe and of some asteroids.

There are dense populations of large impact craters on Umbriel and Oberon very similar to those of the ancient cratered highlands of Earth's Moon, the ancient cratered terrains of Mars, and some of the satellites of Saturn such as Tethys, Dione, and Rhea. These surfaces were probably formed about 4 billion years ago when the planetary bodies were sweeping the debris that remained after the planets had accreted from larger planetesimals. Ariel and Titania have fewer such craters and their surfaces appear to have been molded by impact debris spread throughout the Uranian systems from the formation of the major craters and later swept up by the satellites. These smaller craters appear on all satellites of the system, as they do in many satellites of the other outer planets.

The surfaces of Titania and Ariel may thus be younger surfaces than those of Oberon and Umbriel. On them the original surfaces have been reworked and the old, larger craters have been obliterated. An outstanding question is, why and how did Oberon and Umbriel escape this bombardment?

Of the surface material it is likely that this consists of the dirty ice of comet nuclei and small asteroids. Because of the intense cratering episodes the original surfaces have been reworked, probably several times. The coloration, or rather lack of color, may derive from the radiation environment acting on the various ices present on these worlds. From the bright ray features seen on these worlds we have to assume that the short-period comets captured by Uranus probably caused the impacts that produce these young-looking features by spreading lighter colored subsurface materials over the darkened surface. These comets will tend to concentrate toward Uranus with the result that there are more young-looking impact craters on the inner satellites than the outer satellites.

In addition, it is probable that the major impact of planetesimals on the inner satellites would also have been concentrated by the gravitational field of Uranus. As a consequence the inner satellites may have been broken apart by these collisions and later reaccreted, thus accounting for the lack of big craters on them. Miranda's unusual topography would be explained by the satellite's having been disrupted and later reformed from the pieces.

The dark surfaces of the satellites may result from the effects of radiation because all satellites spend a considerable amount of time in the magnetosphere. When trapped ions in the radiation belts impact the icy surfaces they could split methane ice into carbon and hydrogen. The hydrogen could easily escape into space leaving the carbon behind to blacken the surface. Less than 100 years would be needed to darken a pristine ice surface in this way. Radiation might also act on methane to produce large molecules which could provide another mechanism responsible for darkening the surfaces.

Miranda, the smallest of the larger satellites (figure 4.19), is the most bizarre. Its surface displays scarps, sawtooth terraces, extensional and compres-

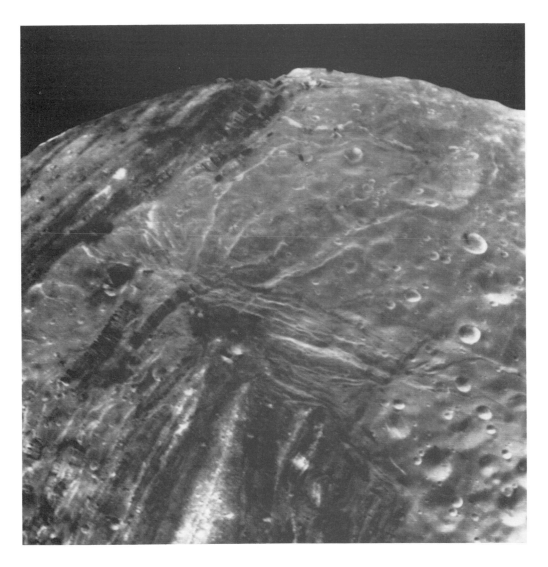

FIGURE 4.19

At 19,000 miles (31,000 km) from the spacecraft the surface of Miranda displays a bewildering array of fractures, grooves, and craters, and areas exhibiting a variety of albedos, some bright, others quite dark. The grooves and troughs reach depths of a few miles and expose materials of different albedos. The bewildering variety of directions of fractures and troughs, and the different densities of cratering, signify that the surface of Miranda has undergone a long and complex evolution. At the top of the image the bite out of the limb is revealed as a wide, deep valley with terraced cliffs. Another wide valley stretches across the middle of this frame at right angles to that running to the limb. At top right is a very large degraded crater. (Photo. NASA/JPL)

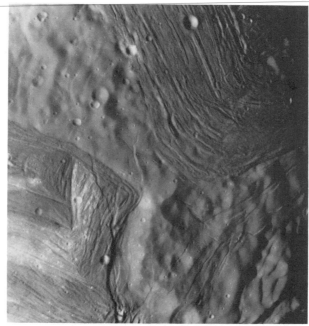

(a)

(b)

FIGURE 4.20

The rolling hill terrain is shown on this image (a) between the dark ovoid and the chevron feature. This is very similar in appearance to the rounded hills of the reprocessed lunar regolith shown in (b). Part of the terrain is marked by shallow and relatively narrow fault valleys. Large craters have been degraded almost to the point of being unrecognizable. Most craters on this terrain are quite small and fresh looking. The intricate faults of the chevron feature are well illustrated in this image including a wide meandering valley cutting across the linear faults, but with the faults still clearly visible on the floor of the valley. This is similar to faults cutting through craters on the planet Mercury as shown in (c). (Photos. NASA)

(c)

sional faults, craters of a wide range of diameters, trenches, and great slabs of terrain. There are geological provinces with very different types of terrain. An old rolling terrain is heavily cratered, has a fairly uninteresting appearance, and does not vary much in albedo from one area to another. A much younger terrain is very complex and exhibits different regional forms. Its albedo varies considerably from place to place.

The old terrain has large impact craters, but they are softly contoured (figure 4.20a), much like areas of the lunar surface (figure 4.20b). On Miranda, this terrain has faults that cut through craters in some places, somewhat like parts of the surface of the planet Mercury (figure 4.20c). These faults tell us that the satellite has been active tectonically since the formation of the ancient terrain. There appears to have been some compression of the old terrain as was also found on Mercury.

Although Miranda is only about 300 miles in diameter—a very small body to develop internal heating—it would appear that the satellite has been molded by tectonics. The necessary heat may have been generated by gravitational torques from Uranus and from the other large satellites. It is unlikely that it was developed in the same way as the internal heating of Io, Jupiter's volcanic inner satellite. Nevertheless heating may have been sufficient on Miranda to cause volcanic flows which are suspected on some regions.

These possible volcanic flows are visible in one of the strange ovoid features (figure 4.21) which have been likened to racetracks because the outer 60-mile (100-km) wide belt wraps around an inner core which is like a rectangle with rounded corners. Two such features were seen on the Miranda image mosaic. They differ somewhat from each other. One has dark bands and a cratered but relatively smooth interior, somewhat resembling a bull's-eye, the other has

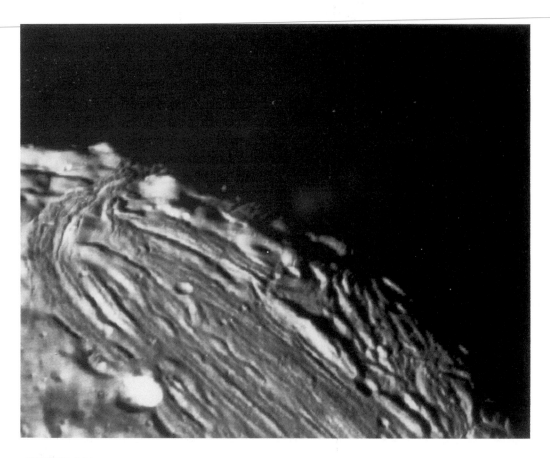

FIGURE 4.21

The dark ovoid consists of an interior of linear ridges and grooves bounded by a racetrack-like system of curvilinear ridges and grooves. Only a relatively few craters are found within the ovoid, indicating that it is relatively young. Next to the ovoid at the bottom left is an area of older cratered terrain with subdued hills. The whole area appears degraded from any pristine sharpness it may once have had. (Photo. NASA/JPL)

complex ridges throughout its inner core and ridges and troughs running along its dark belt. A third strange feature was the chevron which proved to consist of a rectangular structure of faults of different albedos (figure 4.22). It appears to have some features similar to the ridged ovoid feature in that the rectangular center is faulted with intersecting ridges and grooves and this is surrounded by a band which sharply cuts off the ridges. The band is, however, much narrower than the bands surrounding the ovoid features.

It is speculated that all three features represent different stages of a similar geological process, each perhaps caused by the sinking of a mass of rock and the displacement of ices toward the surface. Possibly the result of a breaking apart of an earlier satellite and its subsequent reassembly from odd pieces followed by evolution into a new satellite (figure 4.23). Such a breakup is suported by the fact that there are no really large impact craters on Miranda as would have been expected if it had kept a surface for 4 billion years or so. Miranda may have differentiated normally after its initial formation to give rise to a rocky core with a shell of ices on which there were large impact craters. The catastrophic event that broke up the original satellite into large chunks of ices and rocks orbiting Uranus could have resulted in these resultant satellites having noncircular, inclined orbits. As these masses accreted back to form a new satellite, the present-day Miranda, this reassembled satellite would move in an odd orbit. It would have been subjected to tidal stresses, as the new satellite was urged into the more normal orbit seen today. This process would have internally heated Miranda so that a partial second differentiation of the rocky and icy masses could have taken place.

That initial differentiation of all the large Uranian satellites could have occurred seems more probable when Voyager data were used to derive new estimates of the masses and diameters of the Uranian satellites. The densities calculated from them showed that the satellites could contain more rock than had been thought likely before Voyager for icy satellites of the outer Solar System planets—so much rock that radioactive decay heat might have been sufficient to release enough heat for initial differentiation.

Ariel is about twice the diameter of Miranda. It is still a relatively small world, but it is very different from Miranda (figure 4.24). Its surface shows much tectonic activity but of an entirely different nature. The satellite is pockmarked with craters, but the outstanding features are long rift valleys stretching across the whole of the face of the satellite seen by Voyager. Higher resolution pictures provide views (figure 4.25a, b, and e) that looked like the canyons of Mars (figure 4.25c). Moreover the canyon floors have structure that suggests flow patterns, as though some fluid has smoothed the floors. Even narrower grooves that look as though they have been carved by fluid flow snake along the smooth floors. Almost at the center of one image of Ariel is a "bridge" across a sinuous rille which is reminiscent of "bridges" over terrestrial lava tubes or over the Hyginus Cleft (figure 4.25d) on the Moon.

At the temperature expected on Ariel, water ice behaves like steel. But methane and ammonia ices could have melted and flowed on the surface of the satellite. Carbon monoxide might also have been a liquid on the satellite's surface. But what released energy for these substances to melt and flow? Ariel has no tidal resonances with any other Uranian satellite. There are no tidal stresses to develop heat in its interior. Again it may have been the result of radiogenic heating because of the large proportion of rock in the satellite. Additionally, the tidal forces that acted on a rebuilt Miranda might also have reflected back to Ariel.

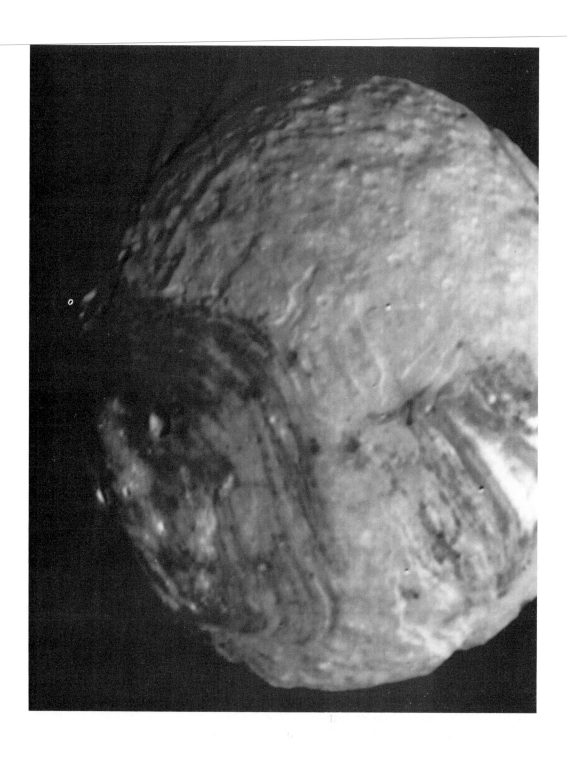

ACCRETION

DIFFERENTIATION

FRAGMENTATION

REACCRETION

FIGURE 4.23
A possible sequence for the evolution of Miranda is shown in this NASA drawing. The satellite accreted normally, differentiated, and was then broken apart by impact of a large planetesimal. The pieces later reaccreted into the jumbled geological jigsaw seen today.

FIGURE 4.22
At a distance of 91,000 miles (147,000 km) a colored mosaic shows a grey surface pockmarked by craters but with many unusual features of grooves and ridges. The sinuous bright marking is revealed as a chevron shape, while the area of dark albedo appears as a banded rectangle with rounded corners. The central part of the rectangle is spotted with bright splotches. On the limb a high mountain is silhouetted against the darkness of space, and not far from it is an enormous chasm. The surface is virtually colorless. The bright chevron feature is revealed as complex parallel ridges forming part of a trapezoidal complex with bright and dark albedos. (See also figure 4.20a). (Photo. NASA/JPL)

FIGURE 4.24
Photographed at 1.56 million miles (2.52 million km) from Voyager, Ariel shows distinct bright areas which reflect nearly 45% of the sunlight falling upon them. These are probably fresh water ice spread over the darkened surface by the impacts of comets on that surface. The south pole of Ariel is slightly off the center of this image. Note the indication of linear faults across the top part of the image and the bright spot at the top limb. (Photo. NASA/JPL)

FIGURE 4.25 (a)
This high resolution mosaic of Ariel results from images taken at a distance of 80,000 miles (130,000 km). Much detail is revealed, including resolution of the linear faults faintly seen on the previous image. Much of the satellite's surface is densely pitted with medium sized craters with only a few large craters, indicating that this hemisphere at least has been molded mainly by secondary impacts. Additionally, internal forces have created many linear faults and scarps and long meandering valleys with flat floors. These valleys are the result of down-dropped fault blocks and are known as graben. The larger fault valleys, near the terminator boundary between day and

(a)

night (on right side of image), have been partly filled with smooth deposits in which there are meandering small channels. Lack of craters on these smooth areas in the valleys and on some areas of the cratered terrain indicate that the smooth material came from a more recent activity than the cratering process. Whether the smaller channels were caused by flow of fluids or by faulting is not clear. However, the major faults most probably resulted from internal expansive forces which stretched and broke the crust. A number of mountains are seen in silhouette against the darkness of space along the limb, and many craters have central peaks like those on Earth's Moon. (Photo. NASA/JPL)

RINGS AND
SATELLITES

131

(b)

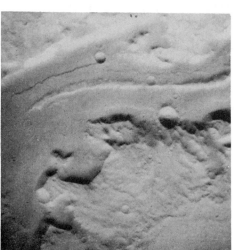

(c)

(d)

FIGURE 4.25 (Continued)

(b) The highest resolution image of the terminator region of Ariel shows the large valley system in more detail. Note that the faults crossing the graben are not visible on the floors of these valleys. Again this suggests that the valleys were filled with material after the faulting process had ended. The sinuous rilles clearly visible on the valley floor in this high resolution image taken from 80,000 miles (130,000 km) obviously formed even later and probably resulted from a fluid flow. Note the similarity to some of the valleys of Mars shown in (c). Almost at the center of the image of Ariel is a bridge over the sinuous rille which is reminiscent of bridges over terrestrial lava tubes or over the lunar Hyginus Cleft (d). However, on the wide valley floor it is positioned as an extension of the fault scarp of a transecting wide valley. If this is its true nature it would have blocked the fluid flow along the sinuous rille.

(e)

FIGURE 4.25 (Continued)
(e) A close-up shows several large craters on Ariel, one with a bright rim and surrounding bright area and a central peak, another with only a few bright spots on its
rim but with a group of three central peaks. Two large craters near the limb have dark
areas on their floors where flows appear to have taken place after the floor was filled.
A smooth triangular-shaped area is shown near the top right corner of the image. This
has only a very few impact craters and appears to be relatively young. On its surface
are a number of ghost craters indicating that the flow is not very thick. Its origin is not
clear.

Further out from Uranus, Umbriel presents another piece of the Uranian
jigsaw. Although about the same size as Ariel and of the same density, Umbriel's surface is startlingly different (figure 4.26). Umbriel is the darkest of the
satellites, and its surface does not change much in albedo from one area to
another. There is no evidence of recent craters, or of impacts that generated
bright ray systems. But there is a strange bright annular feature connected with
the floor of a large crater. Umbriel's surface appears old because it is pockmarked with large craters. The question is what has happened to prevent more
recent impacts from showing up on the surface? One suggestion is that a
volcanic eruption pushed dark material into orbit around Uranus at the radius
of Umbriel's orbit, possibly from Umbriel itself. The satellite would have swept

FIGURE 4.26
An early image of Umbriel taken at a distance of 650,000 miles (1.04 million km) was in color. The surface is characterized by its overall dark coloration and lack of any variations in brightness. The surface is generally grey and colorless and covered with impact craters. A bright ring near the satellite's equator appears unique. The image reproduced here was taken later from a distance of 346,000 miles (557,000 km). It shows evidence of only minor internal activity and a surface molded mainly by impacts. The surface looks very similar to the highland areas of Earth's Moon, the oldest parts of the lunar surface. Strangely, there are virtually no bright young craters as on the other Uranian satellites. A large crater on the terminator has a bright central peak, and the curious bright ring appears to be the wall of a large crater about 90 miles (140 km) across. The nature of this ring and its possible origin are enigmas. (Photo. NASA/JPL)

up the dark material to obliterate any bright areas beneath a thin veneer of carbon-rich deposits.

Then the question becomes how was the volcanic eruption generated? The answer may be that it was not volcanic but an isolated recent impact of an asteroid or a de-iced comet nucleus that generated the material which entered orbit around Uranus. A solution to the mystery must await another mission to Uranus.

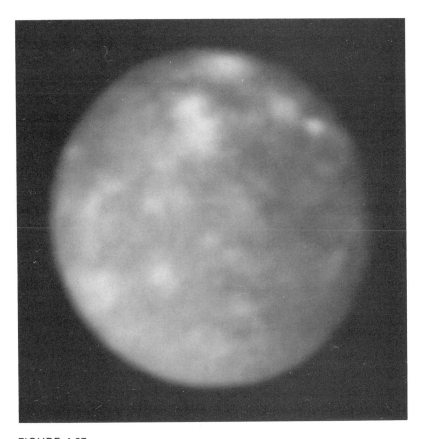

FIGURE 4.27
Titania's surface from a distance of 1.93 million miles (3.1 million km) displays a variety of features, areas of lighter and darker material, and generally a surface that is quite different from the surface of Umbriel. It is very grey and lacks color. At 300,000 miles (500,000 km) the surface shows much detail. Many circular depressions of impact craters are visible, a large proportion of which have bright surrounds and ray systems. The rays are probably associated with the most recent impacts. Linear troughs break the crust in two directions and are probably fault canyons caused by expansion of the satellite's crust. A large impact basin is visible near the top right region of the terminator. It has concentric mountain rings. At the bottom of the image is a double-walled large crater with sharp walls and a light colored floor. On a color image there is evidence of a concentric light band around this crater. (Photo. NASA/ JPL)

Titania, the next satellite out from Uranus is larger than Ariel and Umbriel. Its diameter is 990 miles (1590 km). It has a few large impact basins of 60 to 120 miles (100 to 200 km) diameter and a surface generally covered shoulder-to-shoulder with smaller craters (figure 4.27). These are believed to be secondaries ejected into orbits by the earlier impact of larger bodies. Their subsequent impacts on the surface of Titania probably wiped out the large craters already there.

FIGURE 4.28(a)

A high-resolution image of Titania obtained from a distance of 229,000 miles (369,000 km) shows features as small as 8 miles (13 km) across. Abundant impact craters pockmark the surface. They are of many sizes from big ringed impact basins to tiny craters at the limit of resolution. The large double-walled crater at the bottom of the image is 125 miles (200 km) in diameter. It is cut by a 60-mile (100-km) wide major fault valley. The impact basin near the top of the image is 180 miles (300 km) across and has a complex internal structure and smaller craters that have later destroyed much of the original crater wall. A prominent feature is the system of fault valleys that stretch across the satellite. One is 1000 miles (1600 km) long and 45 miles (75 km) wide. Some of the scarps are very bright, due partly to incident sunlight but perhaps also to young deposits of frost. At least two directions of faulting are apparent.

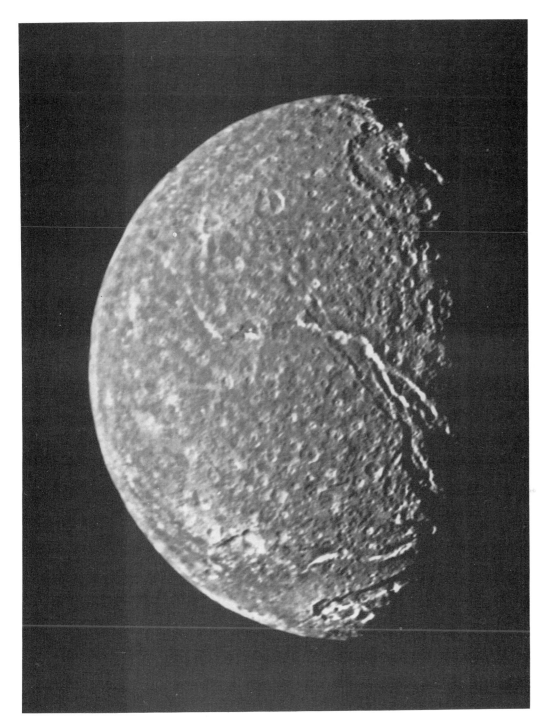

FIGURE 4.28(b)
An enhanced image which reveals more of the surface details showing not only how heavily pockmarked it is with craters, but also the extent of the fault system.

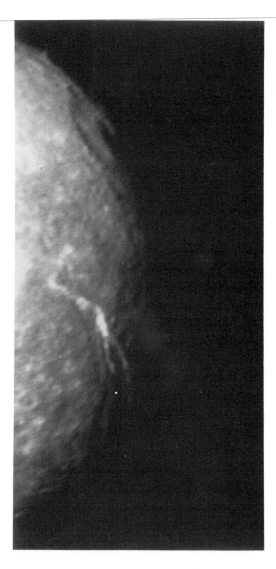

FIGURE 4.28(c)
A blowup of the fault system at the terminator also shows above the faults a large impact basin with at least two mountain chains surrounding it. (Photos. NASA/JPL)

The images suggest that there has been some resurfacing by flows which have given rise to darker, smooth patches. Titania's surface is also marked by many faults which indicate that there have been internal changes molding the surface following the external impact molding. The network of faults has branches, and there is evidence of blocks of surface material dropping to form mile-deep, flat-floored valleys (known as grabens) which have widths between about 12 and 30 miles (20 and 50 km) (figure 4.28). Along several of the cliffs produced by the faults there are streaks of lighter material. Down-dropped blocks are indicative of crustal expansion which might have resulted from the freezing of interior water. Water expands during the last stages of cooling

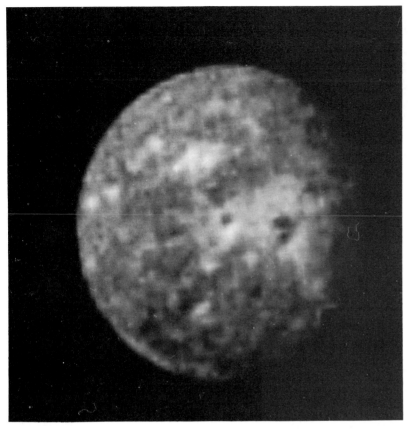

FIGURE 4.29
A long-range picture of Oberon is dominated by a dark spot surrounded by a lighter area. In this image, taken at a distance of 1.72 million miles (2.77 million km), the satellite appears colorless but with areas of darker and lighter materials. (Photo. NASA/JPL)

before it becomes ice, so after the tops of ancient oceans froze into crusts, the water interior would continue to cool and expand on freezing to extend the crust and create the faults cutting across the craters formed by earlier impacts on the icy crust. Because there are few craters on the faults, the faulting process must have occurred after the cratering. Titania also has a few bright ray craters caused, it is believed, by the impact of comets.

Oberon is about the same size and density as Titania but again is different. Oberon is 362,600 miles (583,530 km) from Uranus compared with Titania's orbital radius of 270,900 miles (435,960 km). Because of its greater distance from Uranus, Oberon could not be imaged at as high a resolution as the other satellites with the result that surface details are not so clearly defined. It is nevertheless apparent (figure 4.29) that Oberon has many more large impact

FIGURE 4.30
At 410,000 miles (660,000 km) from Oberon, large impact craters are clearly seen, several with bright rays like those seen on Callisto, one of the Galilean satellites of Jupiter. The dark spot near the center of the disk is revealed as a large impact crater with a bright central peak and dark floor. On the limb a lonely high mountain pokes 4 miles (6 km) above its surrounds. (Photo. NASA/JPL)

craters than Titania and about the same number of smaller craters as has Titania. The general albedos of the two satellites are about the same, but Oberon has more areas of darker material which may indicate a period of fluid flow into and around the craters after the heavy bombardment had ended (figure 4.30). However, some bright ray craters, which are obviously young, also have dark interiors. This may indicate that the outward flow of dark material has occurred sporadically in episodes spread over a long period of time, whenever the conditions were ripe; that is, whenever the outer crust was disrupted. It may be that this dark material contributes to the slightly reddish color of this satellite, unique among the Uranian satellites. Oberon also has a number of faults which again indicate that the satellite has been modified from within as well as from without.

An interesting feature observed on Oberon was a lone, limb mountain poking nearly 4 miles (6.5 km) into space. Otherwise the limb was smoothly circular. This mountain might be the central peak of a large impact basin of which there is some suggestion in the patterns of light and shade on that part of the image.

5
COMPLETING THE GRAND TOUR

In 1811 in the township of St. Lo, Normandy, Urbain Joseph Leverrier was born. In his late teens he became fascinated by the challenge of mathematical astronomy and accepted it as his life's work. Across the English Channel in Launceston, Cornwall, John Couch Adams was born a few years after Leverrier. He also was destined to become a mathematician. During their lives these two men would again double the size of the Solar System in the minds of men.

Following Herschel's discovery of Uranus, astronomers had watched the planet moving slowly year by year in its apparent path through the stars of the Zodiacal constellations. The planet progressed from Gemini, where it had been discovered, into Cancer and then into Virgo. Astronomers became increasingly chagrined as they discovered that Uranus departed more and more from its calculated position. Either the laws of celestial motion discovered by Kepler and Newton did not apply in the extended Solar System or something was perturbing the motion of the newly discovered planet.

The problem was accentuated when astronomers searched earlier records and discovered that Uranus had, in fact, been observed a number of times previously but had not been recognized as another planet. It may have been catalogued as a star by Tycho Brahe in 1589, by Tobias Mayer in 1756, by Flamsteed in 1690 and in 1714, and by P. Charles de Monier several times in the period 1740–1770. There were at least one dozen observations useful in connection with postdiscovery observations for the task of determining the orbit of Uranus.

The problem was that all these observations did not fit together; Uranus appeared to be drifting away from where it should have been when its orbit was calculated, even taking into account the perturbations upon that orbit by the other planets of the Solar System. At first the perturbing effects of comets was thought to explain the discrepancies, but this did not suffice.

Uranus was off course. Why?

The great Prussian astronomer Friedrich Wilhelm Bessel wrote about 1840 to the celebrated naturalist Baron von Humboldt: "The time will come when the problem [of Uranus' motion] will perhaps be solved by the discovery of a new planet, whose elements will first of all be determined by its effect upon Uranus."

By the summer of 1841 John Couch Adams was a young mathematician studying at Cambridge University, England. He had been somewhat a child prodigy in mathematics as applied to astronomy. He first heard of the problem of Uranus' motion from, it is believed, an announcement by the Gottingen Academy of Sciences that a prize was being offered for the best analysis of what might be causing the peculiarities of Uranus' orbital motion. Adams had also read a paper by the British Astronomer Royal, George B. Airy, describing but not explaining the peculiar orbital path of Uranus.

By the time Uranus had moved into the constellation of Libra, Adams, who was still a student at the University of Cambridge, had heard of the irregularities of Uranus' motion, and he had resolved to grapple with the problem as a mathematical challenge. But it was not until he obtained his degree in 1843 that he really concentrated on the task. Adams studied papers of Alexis Bouvard of France, and of Bessel, both of which suggested the existence of another outer planet. Later, he was encouraged by what he had heard through a third party: that European astronomers believed a more distant planet might be perturbing the orbit of Uranus.

Meanwhile, in France, Leverrier was also independently tackling the problem, encouraged by the Director of the Paris Observatory, François J. D. Arago. He did not know Adams, nor did Adams know him. Leverrier, however, also believed that the perturbations were caused by a more distant planet.

Adams made a calculated guess that the trans-Uranian planet would follow what was termed Bode's law, an empirical spacing of planets by distance from the Sun. This would place the unknown perturbing planet at twice Uranus' distance from the Sun. It was a place to start and from it the orbital period of the perturbing planet could then be calculated and a mass calculated that would account for the perturbations observed in Uranus' orbit. Using the details of the planet's position recorded over the years 1754 through 1830, supplied by Airy, Adams continued with the difficult mathematics, aiming to find where the eighth planet would be located at specific dates so that it could be searched for. Within two years he had determined a position where this outer planet should be seen among the stars of the constellation Aquarius, a quarter way around the Zodiac ahead of Uranus.

Leverrier also was working diligently at the problem. He, too, tried to determine the position of the unknown perturbing planet. He, too, used Bode's law to set its distance as twice that of Uranus. Throughout the summer of 1845 the two men zeroed in on the solution. They were able to predict where the eighth planet might be found. In October 1845, Adams sent his predictions to the Astronomer Royal through his professor, James Challis, Director of the Cambridge Observatory, but Airy was out of the country. In several visits to Airy's office, Adams missed the astronomer. When Airy finally saw Adams' paper he was skeptical of both the mathematics and the results. Airy did not place much credence on the calculations of the 26-year-old relatively unknown mathematician. He wrote to Adams asking for answers to a number of questions and expressing doubts about the reliability of Adams' predictions. Annoyed at the doubts concerning his work, Adams left the questions unanswered and ignored Airy's letter, hoping he could convince Challis that the Cambridge University telescope should be used to search for the planet. He was not successful in doing so.

Meanwhile, the astronomer William R. Dawes happened to see Adams' paper and wrote to his friend William Lassell of Liverpool, England, discoverer of two satellites of Uranus, who possessed a powerful reflecting telescope. Dawes asked Lassell to search for the trans-Uranian planet. But Lassell had hurt his leg and was indisposed, and unfortunately Dawes' letter became misplaced so the search was never made.

Leverrier had better luck. He finished his work within a year, when he was 35 years old. He presented his calculations and predictions to the French Academy in November 1845 and again in August 1846, a year after Adams' report to Airy. In due course Airy received a copy of Leverrier's paper and saw that the predictions were the same as those made by Adams almost a year earlier. He sent Leverrier the same questions he had asked Adams, and Leverrier answered them to Airy's satisfaction so that Airy now accepted the possibility of there being another planet. As a result Airy, now more enthusiastic, asked Challis to use the Cambridge University telescope to look for the eighth planet in the constellation Aquarius as predicted by Adams and Leverrier. But Challis did not possess good star maps. He was looking for a very faint object—twice the distance of Uranus and probably not showing a disk—and he needed to know which faint star might indeed be really a planet. He observed and recorded several thousand faint stars, including the trans-Uranian planet but he did not identify it. In one month he accurately measured the positions of more than 3000 stars, but he put off the formidable task of reducing the measurements onto a chart until he had finished his observations. This delay prevented him from being the discoverer of the eighth planet.

In France Leverrier was also enduring disappointments, as French astronomers were slow in starting the search. Frustrated, he wrote to Johann Encke, Director of the Berlin Observatory, urging him to search for the trans-Uranian planet at the position he gave to him. He stated that "it is a remarkable circumstance that there is only one position which can be assigned to the location of the disturbing body." He even suggested that the magnitude of the planet would be about nine, considerably fainter than Uranus.

Encke gave the task to Johann C. Galle, who started observing on the night of September 23, 1846. He was assisted by another astronomer, Heinrich Louis d'Arrest. While Galle peered through the big refracting telescope at the Berlin Observatory and described the configuration of the stars he saw in the region of Aquarius, d'Arrest checked them one by one on a star chart. Fortunately, they had good star maps and after about half-an-hour's observing Galle's comment "There is a star of the eighth magnitude in [such and such] position," d'Arrest immediately exclaimed, "That star is not on the map."

Indeed the faint star-like object was not on the map. The next night Galle observed that the faint star had changed its position slightly. He had found another planet. He wrote to Leverrier. A trans-Uranian planet had been discovered on September 23, 1846.

The size of the Solar System had again been increased enormously.

Back in England, when the news arrived, Challis checked his records and found that he had actually seen the planet a month before Galle, but had thought the object was a star.

Once again naming the planet become a problem. As some English astronomers had wanted to call Uranus Herschel, the French astronomers wanted the new planet to be named Leverrier. Astronomers of other nations again objected strongly, so the tradition of using names from mythology continued. Neptune is the Roman god of the ocean.

In 1980, Charles T. Kowal found that Galileo's records suggest that he might have seen but not recognized Neptune. This was when Galileo was mapping the stars seen around Jupiter. He recorded stars on December 28, 1612 and January 28, 1613, which are not on modern star maps and might have been Neptune.

Joseph J. Lalande also recorded Neptune during his observations on May 8 and 10, 1795, but mistook it for a star in his 50,000 star catalog, *Histoire Celeste Française* (1801). Also in an atlas compiled by K. L. Harding and published in 1822, Neptune appears as a "star" of the eighth magnitude. It is probable that Neptune had been seen as a "star" by many other diligent observers of the heavens.

Neptune is certainly not an object for easy observation. Its angular diameter when closest to Earth at opposition is only 2.5 arcseconds, one tenth the size of Mars at a good opposition of that planet, and two-thirds the size of Uranus. It needs a telescope that can stand a magnification of at least 500 times (i.e., of at least 5-inches aperture) to display an easily recognizable disk.

Neptune travels in its enormous orbit around the Sun at a mean distance of

2.79 billion miles (4.49 billion km), which is just over thirty times Earth's distance from the Sun. It takes 164.79 years to make a complete circuit. Trying to see details on Neptune from Earth is like trying to see details on a quarter from a distance of a mile. At Neptune sunlight is almost 1000 times less intense than at Earth, making the planet and its satellites very faint objects as they reflect this very weak sunlight back toward Earth.

Very little is known about this distant world other than it is almost a twin of Uranus in size. There the resemblance appears to end. Neptune rotates more like the other planets, on an axis nearly vertical to the plane of its orbit; it does not appear to have a system of regular satellites similar to those of the other giant planets.

Neptune's diameter is about 30,200 miles (48,600 km), and it rotates on an axis tilted 28 degrees to the plane of Neptune's orbit. The planet is thought to be warmer than would be expected from incident solar radiation and to emit more heat than it receives from the Sun.

An important aspect of the discovery of Neptune was that the planet had been found as the result of applying mathematical theory—literally discovery at a desk, rather than discovery by observation with a telescope. That discovery ushered in an era in which science became accepted by a wide spectrum of the populace as being a way to solve many problems, replacing antiquated politics and social traditions. It finally dispelled lingering concepts from the Middle Ages and demonstrated that by trying to understand the rules governing the universe, people can make predictions that are later borne out by observations. Indeed, Leverrier enthusiastically predicted that mathematicians would be able to use perturbations of Neptune's orbit to find even more distant planets.

Actually, the orbits and distances assumed by both Adams and Leverrier to predict the position of Neptune were in error. But they had to make assumptions to eliminate some of the variables and thereby reduce the complexities of the mathematical analysis to a point at which it became practical to complete. Much controversy followed as to whether or not the discovery had been a fortuitous accident. It so happened that at the time of the planet's discovery, the orbit assumed by Adams and Leverrier placed Neptune in almost the same position it would have occupied had their assumptions been correct. The mathematicians assumed a somewhat eccentric orbit; actually the orbit was nearly circular. However, between 1800 and 1850 the perturbations on Uranus were at their greatest and could be used to predict Neptune's orbit more easily.

Shortly after the discovery of Neptune, William Lassell, on October 10, 1846, discovered that Neptune has a fairly large satellite revolving in a retrograde but almost circular orbit at a high inclination (158.5 deg.) to the plane of Neptune's orbit. This satellite was named Triton. It is about 2175 miles (3500 km) in diameter, slightly larger than Earth's Moon. The mean distance from Neptune is 220,800 miles (355,000 km), and the orbital period is slightly less than 6 days. In the rest of the Solar System satellites with circular orbits revolve close to the equatorial plane of their primary, whereas satellites in inclined orbits usually have orbits of high eccentricity. Triton is an odd satellite in this respect.

The sky survey undertaken by Gerard Kuiper with the 82-inch McDonald Observatory telescope revealed a second Neptunian satellite in 1948. The satellite is named Nereid. Its orbit is quite different from that of Triton and fits the general pattern more closely; it is more like the outermost group of Jupiter's satellites. It revolves in a very large and eccentric orbit at a mean distance of 3.46 million miles (5.57 million km) from Neptune, taking one Earth year for a complete circuit. The orbit is inclined 27.5 degrees. Nereid is believed to be a captured body, perhaps an asteroid or the nucleus of a large comet. But whether it is a rocky or an icy world we do not yet know. Its diameter is estimated as being only about 200 miles (330 km), but this is a very uncertain dimension.

A third satellite of Neptune, named provisionally 1981N1, has possibly been found. On May 24 1981, Neptune approached close to a faint, uncatalogued star whose light was monitored to probe the space around the planet. Astronomers were searching for evidence of a ring system. The close approach was observed and recorded at two observatory sites in Arizona, and the astronomers found that at both sites the light from the star was interrupted momentarily. This has been interpreted as being caused by a small satellite with a diameter of at least 110 miles (180 km) and, if in the equatorial plane, orbiting at a distance of 47,800 miles (77,000 km). Alternatively, the effect might have been caused by a ring arc, by a part of a ring rather than a specific body.

The satellite system of Neptune is thus a very irregular system compared with those of the other outer planets. Details are given in table 5.1.

Triton's orbital period is 5.88 days, and its rotation is synchronous with its revolution about Neptune. Its orbital motion has been studied mathematically by several researchers to ascertain the physical properties of Triton and Neptune. The relative masses of the two bodies have been determined, and their tidal interactions characterized. The great inclination of Triton's orbit leads to large seasonal changes on the satellite, with large areas of the surface at times denied any solar radiation and thus becoming extremely cold—cold enough to condense atmospheric gases into liquids. For example, nitrogen might be trapped in the cold region and extracted from the atmosphere to form shallow seas of liquid nitrogen. Nitrogen evaporated back into the atmosphere by solar heating would again be extracted in the cold regions and recycled back into the seas.

The size, mass, and composition of Triton are still very uncertain. The effects of tidal forces on the satellite are constrained by its physical nature, and are poorly understood at present. Some scientists' estimates of tidal effects

TABLE 5.1
Known Satellites of Neptune

Name	Diameter miles	Diameter km	Radius miles	Radius km	Orbit Eccentricity (e)	Inclination (degrees)
Triton	2175	3500	220.80	355.00	<0.0005	158.5[a]
Nereid	200	330	3.46	5.57	0.76	27.5
1981NI[b]	370[c]	600[c]	47.80	77.00	[c]	[c]

[a]Unknown or very uncertain.
[b]May, however, be a ring arc and not a satellite.
[c]More than 90 degrees because in retrograde orbit.

indicate that if Triton originally had a very eccentric orbit, it would have changed considerably over a billion years or so; others estimate that the changes would have taken place in millions, rather than billions, of years. If no evidence of tidal heating or surface effects are found on Triton when Voyager scans its surface in 1989, then Triton must be assumed to have followed a near circular orbit, as it does today, for billions of years. This then puts constraints on theories of how Triton became a satellite of Neptune since a captured satellite would be expected to start off with a highly elliptical orbit.

The equatorial plane of Triton, assuming Triton, like Earth's Moon, rotates with one face toward the primary ("spinlocked"), is inclined some 22 degrees to the plane of Neptune's orbit. It precesses in a period of between 600 and 700 years which is combined with Neptune's orbital motion over a period of 166 years. So while the Sun in Earth's skies appears to move relative to Earth through seasons that bring it overhead at noon as far north as 23.5 degrees (Tropic of Cancer) at the summer solstice, and 23.5 degrees south of the equator (Tropic of Capricorn) at the winter solstice, the subsolar latitudes on Triton move from a maximum of 50 degrees north and south of its equator, in a period measured in hundreds of years and with lesser latitude excursions between the extremes. This will have very complicated effects on weather and, indeed, on climate on Triton, even on the composition of the atmosphere at different times. While Earth's Antarctic and Arctic regions can be without sunlight for six months at a stretch, the polar regions of Triton can suffer 50 years or so without any sunlight falling on them. Triton may thus be perhaps the most fascinating world of the outer Solar System, even more bizarre than some of the other strange satellites whose surfaces have been revealed by the Voyagers.

Earlier studies of the orbit of Triton concluded that it could have been reduced from a highly eccentric form to its present almost circular form, and that these tidal actions could have been responsible for interactions with Nereid to produce the latter's highly eccentric orbit. If this tidal friction effect were so, then Triton might have been captured, and an interaction with Pluto, which is discussed in the next chapter, might have been possible also. However, more modern calculations indicate that Triton may have possessed a circular orbit for much longer and that its orbit could not have been decayed by tidal forces from a highly eccentric to a near circular orbit during the lifetime of the Solar System. A current scenario for the Neptune system is that Triton and Nereid are remanent planetesimals which were captured by Neptune. During the capture process any regular satellite system the planet might have had, such as one similar to the system of Uranus, was destroyed. Such a capture process is feasible if at the time of capture there was an extensive nebulosity around the still condensing planet sufficiently dense to provide the drag necessary to change a parabolic or hyperbolic orbit into an elliptical one and thereby lead to capture. However such nebulosity might imply that a regular satellite system could not have formed, or was still in the process of accretion, at the time of capture.

Some modern calculations have shown that after Triton was captured, its orbit would have been changed to a circular orbit within millions of years, and that the tidal effects would have generated heat within Triton and probably caused differentiation during which heavier materials sank to form a core and gases rose to the surface to create an atmosphere. Today, Triton is certainly massive enough and cold enough to hold an atmosphere, but determining if such an atmosphere is present on the satellite is very difficult to accomplish from Earth.

There have been many attempts to calculate the mass of Triton and, when coupled with the range of radii determined for the satellite, they often lead to a wide range of values for its density. However, the densities derived all appear to be fairly high for an icy body in the outer reaches of the Solar System. If these densities are real, then Triton may represent a somewhat different class of body from the large satellites of the other outer planets.

The spectrum of Triton has been explored from Earth and there have been some interesting results. Absorption bands attributable to methane have been observed. Another feature of the spectrum has been attributed to nitrogen, probably present on Triton as a liquid in the form of shallow seas or lakes. Methane is probably present as snow and ice. It has been suggested that because methane is soluble in liquid nitrogen and has the effect of raising the temperature at which liquid nitrogen freezes, those parts of Triton which have a long lasting period of night will develop ices of nitrogen. The nitrogen seas may freeze just as the water seas of Earth freeze in Arctic and Antarctic regions. Indeed, Triton may have permanent extensive polar caps extending almost to mid latitudes. The extent of these caps would be expected to vary considerably and in a cyclic fashion over long periods as Triton's highly inclined orbit precesses around Neptune.

The fact that the surface of Triton reflects more red light than blue light may also indicate the presence of complex organic molecules in the nitrogen oceans and mixed with the water and methane ices. The spectrum also suggests the presence of water ice on Triton. So we might consider the possibility of icebergs of water, methane, and nitrogen floating in extensive shallow oceans of liquid nitrogen intermingled with organic compounds of carbon and hydrogen.

As for the atmosphere, this may be mainly nitrogen varying in surface pressure at different seasons as nitrogen freezes or condenses into the liquid nitrogen oceans. The constituents and pressures within the atmosphere will be determined, it is hoped, by a radio occultation experiment when Voyager flies behind Triton in 1989.

If Triton is uniformly covered by ices the surface temperature will not vary greatly over the seasons, and the atmosphere will always have about the same mass. However, if freezing occurs only at the poles and on high elevations, and sublimation occurs in the equatorial regions, this would cause all volatiles to move from the equatorial regions to become trapped in polar caps. The equatorial regions, with nothing left to absorb heat in sublimation would heat further, while the polar regions would continue to cool. The effect would be for the mass of the atmosphere to change seasonally.

Triton is now approaching the period (around 2006) when an extreme condition will occur and the mass of Triton's atmosphere is expected to increase considerably over what it is today. If Triton's polar caps are sufficiently thin to sublimate entirely during a summer, the other pole would still collect ice, and the mass of the satellite's atmosphere might not change so much. However, as the length of seasons on Triton is very long, the atmospheric mass might still change but not as much as it would if both polar caps survive summer, thereby providing a major source of the volatile constituents of the atmosphere during the whole of the summer period. One thing is for sure—the seasonal behavior of Triton may be unique in the entire Solar System. Voyager should throw light on this intriguing question.

Very much less is known about the physical properties of Nereid, and virtually nothing about 1981N1. If Nereid has been captured, and that seems most likely, it can be an icy world like the nucleus of a comet, or it could be a silicate world like certain asteroids. The satellite 1981N1 may be a body similar to the small satellites discovered by Voyager within the orbit of Miranda, or a satellite similar to Miranda.

In 1982, at a meeting of the American Astronomical Society held in New York City, a team of researchers from Villanova University, headed by E. F. Guinan, announced the discovery of rings around Neptune. The story, however, actually began some years earlier, during an attempt to investigate the Neptunian atmosphere.

On April 7, 1968 Guinan had traveled to New Zealand, where he made photoelectric observations of the occultation by Neptune of a 7.8. magnitude star, BD-17 4388. They were recorded on a strip chart recorder. Unfortunately, the charts were lost in transit back to the United States. Records had also been made on punched cards, but there the data had been sampled at a much lower frequency. Because of the low sampling rate, the cards were put aside.

More than ten years passed and the cards languished in storage. Then they were looked at again and the data were found to be of excellent quality showing a well-defined decrease in brightness before the star was occulted by Neptune. There had been no clouds on the night of the observation so the decrease in intensity of light from the star could not be attributed to effects of the Earth's atmosphere. Moreover, the brightness varied irregularly, so the effect was unlikely to be caused by a single and as yet undiscovered inner satellite of Neptune.

Guinan suggested that if the effect were caused by rings around Neptune, the ring system would be in the equatorial plane of Neptune and at a distance of between 18,675 miles (30,050 km) inner edge and 21,788 miles (35,060 km) outer edge. This is well inside the orbit of 1981N1. The path of the star at the May, 24 1981, close approach as viewed from Arizona did not pass through the area around the planet where the rings might be situated. Indeed, occultation of stars by Neptune are somewhat unusual events. Guinan pointed out that opportunities to check for rings would be presented again by occultations in 1981 and 1983. Unfortunately these occultations did not produce clear evidence of Neptunian rings. Even infrared searches which proved successful in obtaining details of the Uranian rings failed to reveal any Neptunian rings. Occultation data obtained in July, 1984, did, however, again seem to show the presence of rings.

One suggestion is that the rings of Neptune, if any, may be only segments, which would account for their elusive nature. Ring arcs were discovered by Voyager in the ring system of Saturn. Other astronomers have pointed out, however, that ring systems may only be associated with planets that have regular systems of satellites. Since Neptune does not appear to have such a system, the planet might also not possess a ring system. The mission of Voyager to inspect the system at close range has therefore assumed greater importance in the study of ring systems of large planets and their possible origin and evolution.

The arcs of ring material may extend as far out from the planet as 3 planetary radii. The plan for Voyager is to have it pass Neptune within 3.2 planetary radii so that it should be able to detect any rings by imaging them and by radio and possibly optical occultation experiments. However such experiments cannot be defined completely until optical navigation data are obtained as the Voyager spacecraft homes in on the planet. But the communication delay will be 8.2 hours, nearly twice what it was at Uranus.

Seen in a telescope under good conditions, Neptune presents a featureless green disk with darkening toward the limb and suggestions of irregular faint spots. The greenish color results from absorption of red light by methane in the planet's atmosphere. From the 1930s the period of a day on Neptune was accepted as just under 16 hours.

On January 19, 1977, a new "day" was measured for Neptune; studies by Michael J. S. Belton and Sethanne Hayes set the rotational period of Neptune at 22 hours. Over a seven-month study of both Uranus and Neptune, the Kitt Peak astromomers obtained exact measurements of the planets with the 158-inch (4-m) telescope and its echelle spectroscope at the Kitt Peak National Observatory in Arizona. The astronomers recorded the Doppler shift when the spectrograph slit was placed horizontally along the equatorial plane of the planet. The light from the receding limb shifts toward the red, and that from the approaching limb toward the blue. In this manner they were able to measure the rotational speed.

The longer rotational periods for Neptune measured in this way produced speculation that the two outer planets may have evolved differently from Jupiter and Saturn with their shorter rotational periods. Jupiter and Saturn are gas giants, while Uranus and Neptune may be ice giants. The formation of Uranus and Neptune may be more akin to the formation of the terrestrial planets than to Jupiter and Saturn.

A major change in the upper atmosphere on Neptune observed between April 1975 and March 1976 resulted in the reflectance of the planet increasing substantially which, in turn, may be indicative of changing weather patterns on the planet. In this series of observations Neptune was found to have brightened in the infrared by a factor of four. Transient, optically thin clouds appeared to be present in the upper atmosphere of Neptune according to a report by a team headed by Richard Joyce working with the 84-inch (2.1 m) telescope at Kitt Peak National Observatory. The clouds discovered were the first positive evidence of weather systems on Neptune. Astronomers had long assumed that Neptune, being so far from the Sun, did not undergo major atmospheric variations and, indeed, many had used the planet as a standard for albedo measurements of other planets in the Solar System.

In 1984, images obtained with the Las Companas telescope showed bright splotches, believed to be clouds, from which a more accurate rotation period of 17 hours and 50 minutes was derived. Later measurements of the movement of spots across the face of Neptune derived a rotation period of 17.83 hours. Even more recently, variations in the reflectance of the planet in the near infrared have been used to establish a rotation period of 18.4 hours.

In addition to spectroscopy and Doppler shift, photometry, and observation of atmospheric features, other means of calculating Neptune's rotational period include the oblateness of the planet.

This method produced rotation periods between 13.5 and 15.6 hours. However, evidence appeared to be mounting that the two distinct rotation periods of 17.8 and 18.4 hours determined by the other methods were real. If so, this would indicate that Neptune has equatorial winds traveling at high speeds of about 225 mph (360 kph), similar to the high winds on Jupiter. If the winds travel in the same direction as on Jupiter then the true period of Neptune would be the longer period of about 18.2 hours.

The direction of the spin of Neptune on its axis was somewhat uncertain because the Doppler measurements, made very difficult by limb darkening, produced widely varying results, and other measurements of rotation rate did not depend on the direction of rotation. Observation of clouds on images obtained by charge-coupled devices appear to confirm that the direction is prograde; that is, the same direction as the rotation of Jupiter and Saturn.

Hydrogen was first reported on Neptune in 1952, but its abundance was then very uncertain. In 1967 a weak spectral signature of methane was discovered with possibly a larger amount in the atmosphere of Neptune than in that of Uranus. Gaseous ammonia, if present, as hinted by the radio-determined temperature of the planet, would be deep in the atmosphere and difficult to detect from Earth. The 1968 data from the occultation (see above) showed that the atmosphere of Neptune is extensive and composed mainly of hydrogen.

There have been many attempts over half a century to construct models of the interior structure of Neptune. Because the planet is denser than Jupiter and Saturn, its interior is expected to contain a large proportion of heavier material. For example, some models assume there are large amounts of metallic ammonia, others a mixture of methane, ammonia, and water ices. The models are very much dependent upon the size of the planet relative to its mass, since these values determine the density. Prior to the late 1960s the most commonly accepted diameter of Neptune was 27,710 miles (44,600 km). The 1968, data redefined the diameter as 30,760 miles (49,500 km), thus changing the calculated density to 1.66. Earlier models had to be scrapped and new models developed.

The temperature calculated for Neptune's atmosphere on the basis of input of radiation from the Sun is about $-377°$ F ($-227°$ C) compared with $-357°$ F ($-216°$ C) for Uranus and $-16.6°$ F ($-27°$ C) for the Earth. The observed temperature is $-360°$ F ($-218°$ C) for Neptune and $-355°$ F ($-215°$ C) for Uranus. Thus, while Uranus does not emit much more energy than it receives from the Sun, Neptune does. This is in part because Neptune receives less solar energy than does Uranus, but the internally generated energy may be about the same because the planets are so close to being twins in size and probably in composition. If there is enough water, methane, and ammonia (i.e., three times the mass of the rocky core) in the interior of Neptune the planet

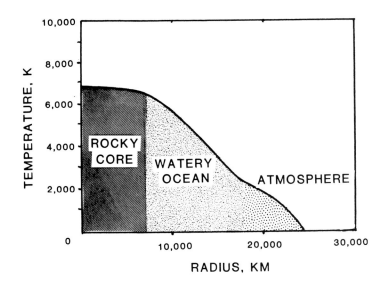

FIGURE 5.1
Graph showing the expected change in temperature from the surface to the core of Neptune which, if true, would mean that the interior ice mantle may be liquid like that of Uranus and could support a dynamo magnetic field.

could have retained enough energy over the its lifetime to radiate energy as it does today. However, if the planet has been formed at a high temperature, it could have reached its present temperature conditions with a lesser amount of water, methane, and ammonia compared with the rocky material.

The total radioactive heating of the core of Neptune (if a rocky core about the size of Earth is assumed) could have raised the temperature of the planet by only a few hundred kelvins. It is unlikely that the present radiation from the planet in excess of that received from the Sun is derived from radioactive heating. The interior of the planet must, however, be subjected to gravitational compression which will be transformed into thermal energy. The interior of the planet could accordingly be raised in temperature to several thousand kelvins by this process. A graph of the expected change of temperature from the interior to the cloud tops of Neptune is shown in figure 5.1. In the model on which the temperature calculations are based, the icy shell is at a high enough temperature to be liquid and to provide the possibility that Neptune, like Uranus, generates its own magnetic field. The presence of the field of Uranus, detected by Voyager, led weight to the interior model of a hot liquid interior shell.

On this basis a cross section of the planet can be derived and compared with those for other bodies of the Solar System, as shown in figure 5.2.

While the atmosphere of Neptune, like that of Uranus, is predominantly hydrogen, it may be that this atmosphere was derived from the decomposition of methane rather than captured from the solar nebula as has been generally assumed. It has been suggested that methane might be pyrolized at high temperature within a planet such as Neptune or Uranus and broken down into carbon and hydrogen. The carbon would fall toward the center of the planet

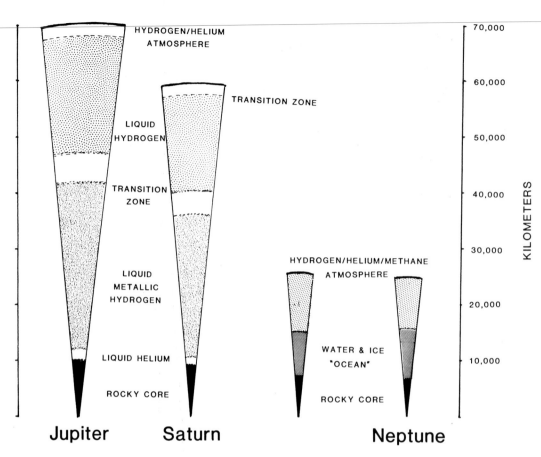

FIGURE 5.2
Cross section of Neptune and the other large planets.

while the hydrogen would rise into the atmosphere. The measure of the hydrogen/ helium ratio is important to checking whether such a process has occurred. From the results of the encounter of Voyager with Uranus, the hydrogen/ helium ratio in the atmosphere of that planet appears to be similar to the ratio for the Sun, so the atmosphere appears to have been derived from the solar nebula and not from the pyrolysis of methane. The same will probably be true for Neptune, but only a measurement of the hydrogen/helium ratio by Voyager will settle this question.

Even before Voyager encountered Uranus, planning was underway at the Jet Propulsion Laboratory for a close approach to Neptune on August 24–25 1989 (figure 5.3). This advanced planning was necessary so that critical resources of the spacecraft could be allocated at Uranus while leaving reserves for the flight to Neptune and beyond.

The plan for Voyager is to have it pass Neptune within 3.2 planetary radii, which is outside the area where it is thought possible for ring arcs to be

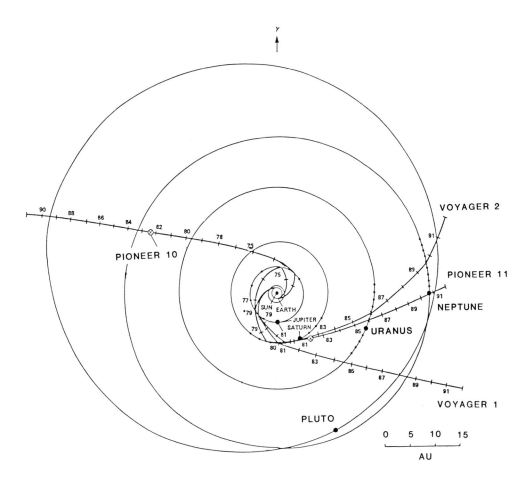

FIGURE 5.3
Paths of the Pioneers and Voyagers through the Solar System showing how Voyager 2 will intercept Neptune in 1989. Tick marks on the trajectories represent years. (NASA)

present. The spacecraft has to avoid passing through rings if it is to survive for a Triton encounter. Voyager will seek to establish if Neptune does, indeed, possess a ring system. The spacecraft should be able to detect any rings of Neptune by imaging them and by radio, and possibly optical, occultation experiments. It should find out if the rings have shepherding satellites and ring arcs and how the ring system, if it exists, differs from those of the other giant planets. It will search for other satellites of Neptune that are invisible from Earth.

However, such experiments cannot be defined completely until optical navigation data are obtained as the Voyager spacecraft homes in on the planet. But the communication delay will be 8.2 hours, nearly twice what it was at Uranus.

At Uranus, the mission had been so successful that there was an adequate supply of consumables aboard the spacecraft for the extended mission. The main operational margin that became important was the availability of electrical power as the radioisotope thermoelectric generators continued to degrade as expected. Telecommunications also posed problems, not only in connection with the eight hour round-trip time but also with the doubling of the range to Earth, which meant that the rate at which data could be transmitted was reduced to one-quarter what it was at Uranus. Additionally, because the intensity of sunlight on the Neptunian system is only about 1/1000th what it is at Earth, long exposures would be needed, and these would smear images unless compensated for.

A multitude of trajectory options were evaluated in order to decide on the one that would optimize the science gathered by this unique mission. After months of debate the scientists decided that the spacecraft should be directed so that it would make a close approach not only to Neptune but also to the giant satellite Triton. The trajectory favored (see figure 5.4) would provide opportunities to gather radio occultation data about the atmospheres of Neptune and of Triton. To do this the spacecraft was directed to fly by Uranus in such a way that it could later dive over the north pole of Neptune, skimming a mere 2200 miles (3500 km) over the planet's cloud tops, and 5 hours later arrive at Triton, passing within 5100 miles (8200 km) of this strange satellite. Another favored trajectory would fly about 3000 miles (4800 km) above the cloud tops and about 24,000 miles (38,600 km) from Triton. The final encounter trajectory will not be decided upon until sometime in 1987 and will be refined further at a later date if needed. Although Triton may have a thin atmosphere of methane and nitrogen, the surface is expected to be visible to the cameras of Voyager. These images of Triton's surface could be very exciting and unusual. They might show surface features and oceans of liquid nitrogen. Triton is tinged with a reddish hue that may be caused by sunlight breaking down methane to produce a smog of hydrocarbon aerosols.

At its encounter, Voyager should provide important and badly needed information about the interrelationship between the atmosphere and the surface of Triton. The radio occultation experiment, coupled with imaging, will determine the radius of Triton with much greater precision than can ever be obtained from Earth-based observations. The flyby will also allow the mass of Triton to be measured with much greater accuracy. Together these two items of information will enable us to calculate the density of the satellite and make more reasonable assessments of what its interior structure may be. A determination of the proportions of rock and ices will provide a basis on which to model the origin and evolution of the satellite, especially when made in conjunction with images of surface features that we hope to obtain.

The size and mass of Triton will also provide a better understanding of what atmospheric gases could be retained—those that cannot reach escape velocity

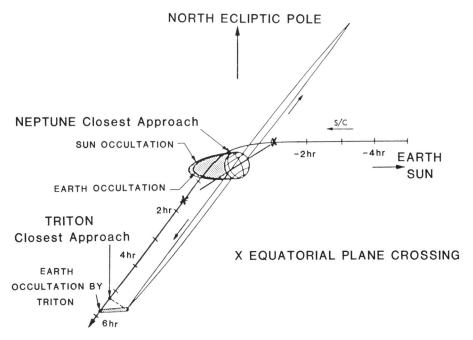

NORTH ECLIPTIC POLE

NEPTUNE Closest Approach

SUN OCCULTATION

EARTH OCCULTATION

S/C

−2hr   −4hr   EARTH
                     SUN

TRITON
Closest Approach

2hr

4hr

X EQUATORIAL PLANE CROSSING

EARTH
OCCULTATION BY
TRITON

6hr

FIGURE 5.4
The path of Voyager through the Neptunian system is over the north pole of the planet, through the planet's shadow, and dipping below the plane of the orbit to encounter the large satellite Triton some six hours later. The spacecraft will also fly through the shadow of Triton. (NASA)

under the conditions on the satellite. An occultation experiment can also establish the scale height of the atmosphere which, coupled with measurements of temperature, allows an estimate to be made of the mean molecular weight and the bulk composition. Then the data from the infrared spectrometer should identify some of the gases in the atmosphere and their relative abundances. This experiment will also provide temperature measurements of regions of the surface and information from which the composition of the surface materials can be inferred.

The imaging experiment should determine if Triton has polar caps, and if so, what are their extent. This is important to estimating the effects of polar caps on the mass of the atmosphere and whether or not this mass will vary greatly over the different seasons experienced on the satellite.

If current models are correct, the polar caps of Triton should be quite large, even on the summer pole, and will remain so through the summer solstice in 2006. If a southern polar cap is not there when Voyager reaches Triton in 1989, then a polar cap will certainly not be lasting until the solstice and a different model of Triton's atmosphere's interractions with the surface will have to be derived.

An interesting surface feature to be sought in the images returned by Voyager will be evidence of seasonal changes or of polar cap shrinking processes and perhaps of cyclical layering as were discovered around the polar caps of Mars.

Voyager's tasks will be essentially to determine if Triton does, indeed, possess an atmosphere and, if so, to try to identify the major constituents and their relative abundances. The distribution of ices and snows between high and low elevations is also important when correlated with surface temperatures. For certain atmospheric models, deep valleys might be expected to have larger deposits of ices and snows, in sharp contrast to Earth, where the ices and snows are concentrated on high elevations. Voyager should be able to establish whether this is true or not and further refine our views of what may be bizarre climatic conditions on this outpost world of the Solar System.

As for Neptune itself, the Voyager flyby will search for cloud patterns that will enable the rotation rate of regions of the atmosphere to be established so that zonal flow patterns, if they exist as expected, can be charted. The magnetosphere will be sought and the magnetic field measured. If the field is offset from the planet's axis of rotation, its wobbling will allow the rotation rate of the interior to be established and compared with the atmospheric rotation rate. Aurorae, airglow, and electroglow will be searched for. The bulk composition of the atmosphere and its constituents and temperature at various levels will be investigated. The hydrogen/helium ratio, the heat balance, the ionosphere, and so on as at Uranus, will also be investigated.

With a successful encounter the physical details of Neptune, its rings, its satellites, and its environment can be compared with those of Uranus. Our understanding of the outer giants of our Solar System will receive an impetus which will keep theoretical astronomers busy for decades in the search for explanations of how the Solar System and its planets were formed.

6
GIANTS AND DWARFS IN THE OUTER DARKNESS

At Lowell Observatory on pine-clad Mars Hill, overlooking the town of Flagstaff, Arizona, is a telescope dome that is small, and somewhat inconspicuous, compared with others in the Flagstaff area. From this location among the pine trees a young astronomer would bring this observatory into a unique club that so far consisted of locations at 17 King Street, Bath, and the Observatory of Berlin. Here at Flagstaff the boundaries of the Solar System would again be extended by the discovery of another outer planet.

Existence of a trans-Neptunian planet was probably first suggested in a paper by Camille Flammarion published in November 1879. It was based on the aphelia of short-period comets. Several of these were known to be associated with Jupiter, Saturn, Uranus, and Neptune, but other comets had aphelia far beyond Neptune. Flammarion suggested that they were associated with a large planet orbiting the Sun at a distance of 4 billion miles (6.5 billion km).

In the United States, David P. Todd, and in Scotland, George Forbes, independently sought to calculate the position of a trans-Neptunian planet. The former arrived at an orbital period of 375 years on the basis of Uranus' deviations. Forbes, however, worked with the comet theory of Flammarion and suggested that there were two trans-Neptunian bodies; the innermost would have a period of 1000 years, was larger than Jupiter, and would be found in Libra. Needless to say, photographic searches did not reveal these planets.

A dedicated aspect of exploration of the outer Solar System began at the end of the nineteenth century when Percival Lowell, after hearing of Schiaparelli's "discovery" in 1877 of channels on Mars, abandoned a promising career in his family's business to concentrate all his efforts on astronomical observations. He was rich enough to establish his own observatory at the verdant Flagstaff "oasis" in the great western desert of Arizona, where he could have good atmospheric conditions for observations. While Lowell's prime interest was Mars, he also became intrigued by the discrepancies in the motion of Uranus that could not be fully accounted for by the presence of Neptune.

Lowell had excelled in mathematics at Harvard and early in the 1900s he also used the aphelia of meteor swarms to suggest an unknown planet located at 45 astronomical units, 4.2 billion miles (6.8 billion km) from the Sun. Later he calculated an orbit for the distant planet based on perturbations of both Uranus and Neptune. He called this body Planet X, and said it would be found orbiting the Sun at a mean distance of 4 billion miles (6.4 billion km) in a period of about 280 years and of mass between six and seven times that of Earth to account for the observed peturbations. He stated that it must be a small planet that would be very difficult to observe at such a great distance, but that it could be located in Gemini or in Sagittarius. He favored Gemini as the more likely position.

In 1905 he started his search for Planet X using a 5-inch aperture telescope coupled with a camera. His aim was to photograph regions of the sky where he believed the planet should be and later compare each photograph with another of the same area taken at a different time. If a faint star had moved between the two photographs, he would recognize it as the planet he was seeking. He would be able to differentiate Planet X from asteroids because the latter would move a greater distance between two pictures. With the long exposures Lowell was using to record faint stars, some asteroids even made streaks and could be identified as such immediately.

In 1915 Lowell, after refining his calculations, confirmed that Planet X would be somewhere in the constellation Gemini. But he still could not find it. Lowell died in 1916, his search for Planet X still unsuccessful—or so the disappointed astronomer believed. Ironically, the planet had been recorded on his photos twice in 1915, but no one at Lowell Observatory recognized the very faint "star" as the sought-for planet.

Another famous American astronomer, William H. Pickering of Harvard Observatory, was also searching for the ninth planet which he called Planet O. He, too, had decided from his calculations (which he had begun about 1910) that Gemini was the likely place to find it. He had used the motions of Neptune as well as Uranus to calculate an orbit and a position for Planet O. In December 1919, he asked an astronomer, Milton Humason, at Mt. Wilson Observatory overlooking Los Angeles, to make a photographic search. Again, the search was reported as being unsuccessful; again, Pickering did not know that pictures had actually been taken of the area of Gemini where Planet O was, and that its image had in fact been recorded on several plates. Pickering concluded there was no such planet beyond Neptune, and turned his attention to other astronomical research.

The search at Flagstaff for Planet X ended for nearly a decade after Lowell's death because of legal problems with his estate. Then it started again. This time it was by a young amateur astronomer who had joined the observatory's staff in January 1929. Clyde W. Tombaugh had built his own telescopes and had been invited by the observatory's director, Vesto M. Slipher, to be a photographic assistant working with a new 13-inch aperture telescope/camera instrument capable of photographing very faint stars. Slipher had decided to continue Lowell's search for Planet X with the new instrument. To begin with, Tombaugh had other duties at the observatory, but gradually he became engrossed in the search for the elusive planet; exposing photographic plates at night and examining them carefully during daytime. He used a blink microscope, first introduced by Lowell before his death, to examine the plates (figure 6.1). The instrument showed first one plate and then the companion in rapid succession. Any "star" that moved would be seen to oscillate as the plates flashed on and off alternately. This was an enormous improvement over Lowell's earlier method of superimposing one plate over another and inspecting the pair with a magnifying glass, looking for any faint star that did not match exactly on the two plates.

The search continued all through the cold winter of 1929 and into 1930. Millions of star images were compared with the aid of the blink microscope. Where the Milky Way passes through Gemini the going became tougher and tougher as the numbers of stars on each plate increased enormously. Inspection, even with the blink microscope, fell behind the exposing and developing of the photographic plates. The work was slow and wearisome, but it had to be performed meticulously. Again, as with Herschel and the discovery of Uranus, Tombaugh was the right man for the task. He was persistent and he was accurate in his work. His comparison on February 18, 1930 of two plates obtained on January 23 and 29, 1930 finally revealed an object that might have been the elusive Planet X. He checked an earlier plate taken on January 21 and again saw the "star" that moved.

The cautious Lowell astronomers did not announce the discovery until the motion of the distant object had been further checked. Subsequently the object moved just as a planet should, but inspections with large telescopes did not reveal a disk. Planet X was not only far away, it was also small—much smaller than either Pickering or Lowell had calculated. The discovery was announced on March 13, Percival Lowell's birthday, and the same day on which Herschel had discovered Uranus 149 years previously.

What to name the new planet? Some people suggested it should be called Lowell, but the name Pluto, reputed to have been suggested by the daughter of

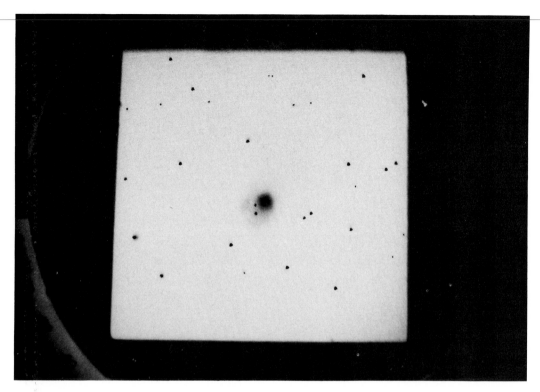

FIGURE 6.1

Blink microscope at Lowell Observatory now used to search for stars with high proper motions. This instrument was used by Tombaugh to compare photographic plates taken a few days apart to look for faint stars that had moved and therefore might be the planet he was searching for. (Photo by Author)

an English astronomer, was finally adopted. Not only did the name Pluto continue the tradition of names from Greco-Roman mythology—Pluto was the son of Saturn and ruler of the dark underworld—but also the first two letters were the initials of Percival Lowell.

Pluto is quite different from the other outer planets. It is much smaller than any other outer planet, and it follows an orbit that at times (1979 through 2000) brings it closer to the Sun than Neptune (figure 6.2). The orbit is not only very eccentric but also highly inclined to the plane of Earth's orbit around the Sun—the ecliptic plane (figure 6.3). All other planets have orbits that are close to the ecliptic plane.

At its greatest distance (aphelion) from the Sun, Pluto is 4.6 billion miles (7.4 billion km) away. At its closest it is 2.8 billion miles (4.5 billion km) distant. This perihelion will be reached in 1989. The orbital period is 248 years.

Pluto appeared to have a long day (6.39 Earth days), as determined by a

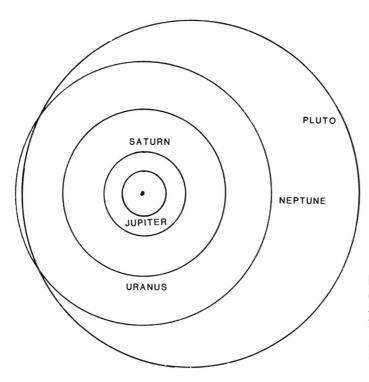

FIGURE 6.2
Orbit of Pluto seen in plan view showing how the planet for a short part of its orbit travels within the orbit of Neptune.

regular fluctuation in the brightness of the planet. The diameter of Pluto has not, however, been determined with any great degree of accuracy. Its mass, too, is very uncertain, so we have no understanding of its interior structure. Without a flyby spacecraft mission to Pluto we are likely to continue in ignorance of this small, remote world. The low mass and low temperature expected of Pluto suggested that it is unlikely to have a significant atmosphere. Water, carbon monoxide, methane, and ammonia would probably all lie frozen on the planet's surface. Hydrogen and helium, still gaseous at Pluto's temperatures, could possess sufficient energy to overcome Pluto's small gravity and escape into space.

Although the orbits of Neptune and Pluto cross, mathematical analyses of the orbits and mutual perturbations of the planets for 120,000 years by C. J. Cohen and E. C. Hubbard of U.S. Naval Weapons Laboratory in 1964 suggest that the two planets can never collide, and that Pluto's orbit is remarkably stable.

The discovery of Pluto raised some very important questions and by no means solved the peculiarities of the motions of Uranus and Neptune. First of all, astronomers had postulated that in order for a planet beyond Neptune to have influenced the paths of Uranus and Neptune, that trans-Neptunian planet would have to be quite massive; Lowell placed it as being at least 6 times Earth's mass, and Pickering said at least twice. Other astronomers suggested that it would have to be a massive gas giant like Jupiter and Saturn.

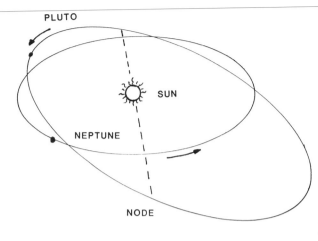

FIGURE 6.3
Oblique view of the orbits of
Pluto and Neptune showing the
high inclination of the outermost
planet.

But Pluto most certainly was not a gas giant. It was not as distant from the Sun as had been expected. Had it been a planet with sufficient mass to account for the perturbations it would have appeared much brighter. Actually it is an extremely faint object as seen from Earth—1000 times fainter than Neptune. On the basis of its brightness and on an assumed albedo, its diameter was originally estimated as being smaller than Earth's, possibly only about 6500 miles (10,460 km). In order for such a small planet to perturb its much bigger neighbors so much, it would have had to be quite massive which, in turn, required that it must be unusually dense compared with any other planet. The problem was compounded when the diameter, following more accurate work, had to be revised downward to only 3600 miles (5800 km), about the size of Mars. To affect the other planets as was observed, an object of that size would have had to be at least fifty times more dense than water.

In 1968, R. L. Duncombe, W. J. Klepcyznski, and P. L. Seidelmann of the U.S. Nautical Almanac Office made some in-depth recalculations of the relationship between the mass, diameter, and density of Pluto. They showed that from the observed perturbations of Neptune the mass of Pluto should be about 0.18 that of Earth. If the density were the same as that of Earth, Pluto's diameter would be 4380 miles (7200 km), whereas for the diameter usually adapted at that time (3800 miles, or 6400 km), Pluto's density would have to be 1.4 times Earth's.

In 1976 the accepted size of Pluto had to take another jump downward. Dale P. Cruikshank, Carl B. Pilcher, and David Morrison of the University of Hawaii identified methane on Pluto by analyzing infrared light from the planet. If Pluto was, indeed, covered with methane ice it would be expected to have a relatively high albedo, which would mean that its diameter might be no bigger than Earth's Moon. Such a small body could not affect Uranus and Neptune in any significant way. (At that time, of course, it was not known that the Uranian satellites, despite having methane identified on their surfaces, had relatively dark surfaces.)

One possibility which had been suggested in the 1950s was that Pluto was found fortuitously. It was not the planet perturbing Uranus and Neptune. A giant planet further out in the Solar System still remained to be discovered. In this scenario Pluto was suggested as the largest of another group of asteroids orbiting generally between Neptune and a large outer gas giant and possibly other smaller planets as yet undiscovered and possibly beyond the capability of Earth-based instruments to find.

Because the mass of Pluto could not be calculated, yet another theory was aired—that Pluto was a special type of high-density body, the dark, burnt-out shell of a companion star to the Sun, possibly covered with a smooth icy sheet of condensed gases. Like a smooth, black, snooker ball, Pluto would reflect only a small spot of light which would lead to underestimating its diameter when observing it from Earth. Such a body could conceivably be 50,000 miles (80,500 km) in diameter and perfectly adequate to tug at Uranus and Neptune so as to perturb their paths around the Sun.

Much of this speculation about size was brought to an end when in 1978 James Christy of the Naval Observatory was examining photographic records to calculate an improved ephemeris for Pluto. He noted that the image of the planet was not a regular sphere. There was a part that bulged. Was this a defect in the photographic emulsion? He checked other plates. The bulge was real enough. He decided that the bulge was caused by the presence of a satellite orbiting very close to Pluto. This satellite was named Charon, after the mythological ferryman who carried the spirits of the dead across the River Styx. But some astronomers did not accept this explanation for the bulge on Pluto and suggested that the irregularity was caused by a feature on the planet, such as a high mountain. This viewpoint was supported by the fact that the bulge appeared to revolve around Pluto in the same period that Pluto rotated on its axis, as determined from variations in the apparent brightness of the planet, but the idea of such a mountain was never accepted as a serious possibility.

The identity of Charon as a true satellite of Pluto was confirmed in 1985 when a rare alignment of Pluto and Charon occurred. It caused the planet and its satellite to eclipse one another approximately every three Earth days. The series of eclipses occur every 124 years and last for five or six years only. Each time Charon passes between Pluto and Earth, a part of the surface of Pluto is blocked by Charon from view (figure 6.4). The light reflected to Earth from the Pluto–Charon system is partially reduced. And when Charon moves behind Pluto it is completely eclipsed. Observations of these eclipses proved that Charon is a real separate body and not a mountain on Pluto. Moreover the durations and changes in brightness enabled astronomers to refine estimates of the masses, diameters, and densities of both Pluto and Charon.

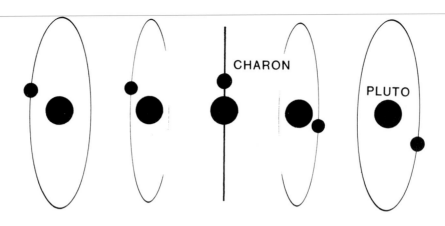

1996   LAST ECLIPSE 1988 FIRST ECLIPSE 1980
1991   1985

FIGURE 6.4
Every 124 years Pluto and its satellite Charon are oriented so that they eclipse each other as seen from Earth. This diagram shows the current series of eclipses.

The first observation of the eclipse sequence was by Edward Tedesco of the Jet Propulsion Laboratory at Palomar Observatory on January 16, 1985. Many other eclipses were later seen by other astronomers. The observations showed that the combined brightness of Pluto and Charon diminished by 4 percent and that this lasted for two hours as Charon passed between Pluto and Earth. This change was superimposed on a 30 percent brightness change that occurs every 6.39 days because of the rotation of Pluto, which alternately exposes bright and dark areas of its surface to Earth.

Today's viewpoint of Pluto and Charon still retains many mysteries. We accept a Pluto of between 1300 and 1500 miles (2100 and 2400 km) in diameter orbiting the Sun at an average distance of 39.44 times Earth's mean distance from the Sun—3.7 billion miles (6 billion km). A year on Pluto is 247.7 Earth years. The orbit is highly eccentric (0.25) and highly inclined (17.2°) to the plane of the ecliptic. The mass of Pluto, determined from its orbital pull on Charon, is 0.0021 that of Earth, making its density between 0.5 to 1.4 times that of water. Its interior probably consists of a small rocky core surrounded by a deep mantle of water ice topped with a crust of methane ice (figure 6.5). An important discovery following the discovery of Charon was that its orbit shows that Pluto is, like Uranus, tipped so that it spins on its side. The axis of rotation of Pluto is tilted 118 degrees and the planet rotates on this axis once each 6.3875 days. Its surface temperature is very low, averaging about −370° F (−223° C), but varying probably as much as 10 degrees above and below the mean value as Pluto moves between perihelion and aphelion.

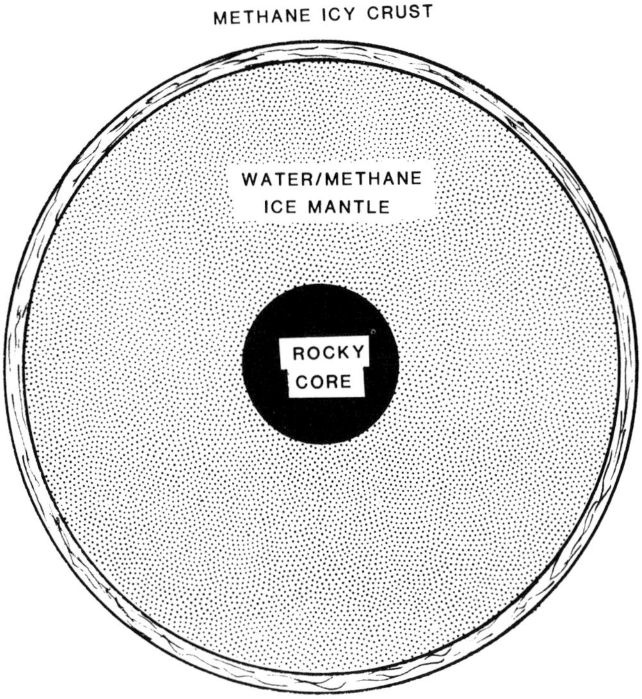

METHANE ICY CRUST

FIGURE 6.5
Current concept of the interior of Pluto—a small rocky core surrounded by a deep mantle of water ice and a thin crust of methane ice.

WATER/METHANE ICE MANTLE

ROCKY CORE

Charon is the only known satellite of Pluto around which it revolves in an almost circular orbit at a distance of 12,055 miles (19,400 km) from Pluto. The orbital period is the same as the revolution of Pluto; namely, 6.3875 days. Its diameter may be about 750 miles (1200 km), its mass is estimated as 0.0002 that of Earth.

Both Charon and Pluto are locked in rotation; they turn their faces continually one to the other. From a viewpoint on Charon where Pluto can be seen, the planet remains fixed in the sky. The same applies to a view of Charon from Pluto.

One thing is certain about this mysterious pair; Pluto and Charon are not the Planet X that had been searched for to account for the discrepancies in the paths of Uranus and Neptune. As with Neptune, the discovery of Pluto seems to have resulted from a fortuitous set of circumstances. It was an object that just happened to be close to a position calculated for a hypothetical planet. The painstaking methodical inspection of this area by Tombaugh led to discovery of the planet, just as Herschel had discovered his planet by methodical examination of the heavens.

In 1933, soon after Pluto's discovery, a Japanese astronomer, Issei Yamamoto, speculated that Pluto is an escaped satellite of Neptune rather than a planet. This theme was developed by R. A. Lyttleton of Cambridge University, who showed that a close encounter between Pluto and Triton could have ejected Pluto from the Neptunian system and pushed Triton into its peculiar orbit. Later still Gerard P. Kuiper suggested that Pluto originated from the planetesimals that were forming the Neptunian system; interactions between these bodies flung Triton into its retrograde orbit and thrust Pluto out of the system.

The idea of the Triton–Pluto interaction has been more recently shown to be impractical. The energy and momentum exchanges in any plausible interaction between the two bodies would have flung Pluto out of the Solar System. Moreover, a twin system such as Pluto and Charon could not have existed in the Neptunian system. Tidal friction would have caused the two bodies to merge into one.

Another interesting possibility is that Neptune originally had a regular system of satellites similar to those of the other large outer planets. A large planetesimal then passed through the system and caused the disruption which lead to Triton's retrograde and highly inclined orbit, while Pluto was freed from Neptune's system but not thrust out of the Solar System. The large planetesimal might still be orbiting in the far reaches of the Solar System.

Today, however, it seems most likely that Pluto and Charon are asteroid-sized planetesimals somewhat akin to the recently discovered Chiron, and Triton and Nereid are similar planetesimals that were captured by Neptune. Such a capture can be achieved if Neptune had at the time of capture a large envelope of gas to slow the speeding planetesimals into closed orbits around the planet.

It now appears that Triton is more massive than Pluto, and is quite a different type of body. If this is true, its greater mass would allow it to hold onto a number of volatiles that would escape from Pluto. The high obliquity of Pluto's orbit will cause any volatiles to migrate from pole to pole over the planet's year. Pluto's atmosphere probably consists of nitrogen, carbon monoxide, and methane, with perhaps some oxygen and other trace gases, probably similar to Triton's atmosphere (which will be investigated and definitized by Voyager) but containing fewer lightweight gases. While on the part of its orbit beyond Neptune, Pluto will be cold enough for most volatiles to freeze and considerably deplete any atmospheric mass. But Pluto reaches a perihelion closer to the Sun than does Triton, so it is expected to be slightly warmer than Triton for part of the Plutonian year, which might produce an atmosphere at that part of its orbit. Pluto also has the peculiar seasonal effect that winter and summer solstices occur when the poles are pointing toward the Sun. Thus alternate poles, rather than equatorial regions, receive the maximum input of solar energy.

The search for most distant planets has continued. For 13 years after his discovery of Pluto Tombaugh continued his photo-mapping (with the 13-inch telescope at Flagstaff) of the region north and south of the ecliptic. He inspected nearly 100 million stars down to the 17th magnitude, but although many spurious images and asteroids were encountered, no evidence of another planet was uncovered by this laborious search.

As pointed out many times a possible clue to the existence of other planets beyond Pluto is found in cometary orbits. Most known comets follow closed orbits around the Sun, but predominantly orbits of very great eccentricity. The ellipses traced by comets extend into the outer reaches of the Solar System. Some comets have such highly elliptical orbits that they are difficult to differentiate from parabolas. Others follow ellipses which have aphelia about at the distances of the orbits of the outer planets. About fifty comets have their aphelia at the distance of Jupiter from the Sun; presumably they are comets that have been "captured" by Jupiter. Saturn appears to control about ten comets, and Uranus and Neptune have small retinues of cometary vassals. There are also groups of comets that reach out at their aphelia to about 7 billion miles (11 billion km) from the Sun, which is about the distance expected of a tenth planet if it followed the approximate relationship of planetary bodies established empirically by J. D. Titius and expanded by Johann E. Bode (director of Berlin Observatory) in 1772 for the then known planets. Subsequent discoveries of asteroids (between Mars and Jupiter), of Uranus, and of Neptune/Pluto (as a group), approximated the rule which states that the distances of the planets are in an approximate geometrical progression.

Following this reasoning, a Soviet announcement in 1975 claimed that two planets might be present according to an analysis of cometary orbits made by the Institute of Theoretical Astronomy, Leningrad. This study found a grouping at 5 billion miles (8 billion km) and another at 9 billion miles (14.5 billion km).

At such distances a planet of normal size would be extremely difficult to detect from Earth. From those distances the Sun would appear as a bright star and the amount of sunlight reflected by a planet far beyond Pluto would be negligible compared with the other planets. The planet's surface would be extremely cold with all atmospheric gases solidified. The time taken for a complete revolution around the Sun would have to exceed 500 years.

Does such a distant planet exist? We have no way of knowing, but we do know that Neptune and Uranus have perturbed orbits which cannot be caused by Pluto alone.

There have been other suggestions. Perhaps dark stars approach close to the Solar System and disturb the planets, or perhaps the Sun is a bright companion to a group of burnt-out stars. Or perhaps a large burnt out star, following a highly elliptical orbit around the Sun, periodically comes close enough to disrupt comets and send them hurtling sunward.

All these are speculative theories. The dark elliptical solar companion has been named Nemesis because its perturbations on comets have been suggested as the cause of mass extinctions of life on Earth, as observed by analysis of rock layers. Some years ago paleontologists at the University of Chicago suggested that such mass extinctions occur every 26 million years, and several other researchers have claimed to show that cratering episodes on Earth follow a cycle of roughly 30 million years. Since impacts of Earth-orbit-crossing asteroids would be random, the periodic nature of meteor impacts and mass extinctions demands some cyclical phenomenon. A large planetary body moving in an orbit reaching to the Oort cloud of comets, possibly in a highly inclined orbit, would dislodge comets each time the planetary body passed through the aphelion of its orbit.

Other very detailed studies have been made on the perturbations of Uranus and Neptune and led to predictions where the large other planet might be found.

At the U.S. Naval Observatory, Washington, D.C., Robert Harrington and Tom Van Flandern used different methods to arrive at a common conclusion. Van Flandern programmed a computer model to show what would be needed for a planet to cause the orbits of Uranus and Neptune to agree with observations, while Harrington developed a computer model to show how a planet might have encountered Neptune and, interfering with its systems of satellites, caused Pluto and Charon to be flung into an independent orbit and Triton into a highly inclined retrograde orbit. Strangely, the planet in both instances turned out to be about the same; a mean distance of 7 billion miles (11 billion km) from the Sun, and about 3 to 4 times the size of Earth. The orbit would be elliptical, making its closest approach to the Sun at 4 billion miles (6.5 billion km) while its greatest distance would be at 10 billion miles (16 billion km).

For over a decade John. P. Bagby in Southern California has investigated theoretical reasons for the existence of the tenth planet and correlated some of the calculated positions with observations of infrared objects with large proper motions. He has concluded that a planet of quite large mass, about one-fiftieth that of the Sun, orbits the Sun at approximately 9.3 billion miles (15 million km). Later he suggested that a mass even greater than the Sun might be orbiting out there, which tied in with the theory of Nemesis suggested by Luis W. Alverez in 1975. Bagby also suggested that the mass causing the perturbations might not be concentrated in one body, but in several bodies having different orbits, some with periods measured in thousands of years.

At a meeting in the summer of 1982, scientists discussed the perturbations of Uranus and Neptune in depth and speculated about the causes. Some argued for an unseen black hole or a neutron star traveling along a path that has brought it close enough to the Solar System to disturb the outer planets. Others pointed out that black holes are X-ray emitters and no such source of X-rays close to the Solar System has been observed. Another suggestion was that a brown dwarf star may be orbiting the Sun, invisible from Earth but possibly detectable by infrared telescopes placed in space. There are many binary stars in the universe, so the Sun might be expected to have a companion also. A brown dwarf might be a very large planet that was not quite large enough for nuclear reactions to start in its interior. The first infrared telescope to be placed in Earth orbit did not, however, produce evidence of such a star.

John D. Anderson of the Jet Propulsion Laboratory, pointed out in 1982 that there were three possibilities for a planet beyond Neptune/Pluto. It could be a giant distant planet with a dark surface, reflecting so little sunlight back into the inner Solar System that it is extremely difficult to find optically. It could be a dark star companion to our Sun, very much further out than a planet and possibly 50 billion miles away. It could be a black hole, much more massive than the Sun and at a distance of 100 billion miles. It was suggested that the two spacecraft, Pioneer 10 and Pioneer 11 that were heading out of the Solar System, be carefully tracked to check if their paths were changed by some large mass. The two spacecraft are flying in opposite directions out of the Solar system. Six years of tracking did not produce any definite results but established limits on the mass and distance of such a trans-Neptunian object.

Anderson points out that during the 100 years between 1810 and 1910, when measurement techniques were comparable with those made with modern instruments, the evidence for an additional solar system object is almost unassailable. The big mystery is that the extremely sensitive measurements of perturbations of the paths of the Pioneers have been negative. Their paths were not perturbed. Checking back on measurements of perturbations to Uranus and Neptune after 1910, Anderson found that many of these were negative, too. The best explanation is that the perturbing planet is following a highly elliptical orbit and it perturbed Uranus and Neptune during the seventeenth, eighteenth, and nineteenth centuries and into the beginning of this century, when close to perihelion, but it is now moving out toward aphelion. If the tenth planet is moving on a very highly inclined orbit, it would fit the data even better.

Mass extinctions of life forms on Earth have been cited as evidence of catastrophic events such as the impact of an asteroid or a comet with the Earth. First evidence for such an impact was presented in 1975 by Dr. Walter Alvarez who cited deposits of iridium in an inch-thick layer of clay separating limestones laid down about the time of the transition from the Cretaceous to the Tertiary periods of Earth's geological history. Iridium comes from meteorites which accumulate it on Earth only gradually. The iridium in the clay suggested that a major impact had occurred, but critics contended that the iridium may have accumulated from volcanics bringing the rare element up from the Earth's mantle. The iridium-rich layer was found elsewhere on Earth at the Cretaceous-Tertiary boundary which seems to imply a world-wide rather than a local deposition.

A large asteroid or a comet nucleus hitting the Earth would cause enormous dust clouds to enter Earth's atmosphere which would cut off sunlight and produce the equivalent of the so-called nuclear winter, halting plant growth and leading to the extinction of species that depend on plants for their food. In turn, carnivores would die as they, too, would be without food. The extinction of the dinosaurs occurred in common with many other life forms some 60 million years ago; but there have been many similar extinctions according to the tale told by the rocks. The most recent occurred about 8000 years ago, when the mastodons and mammoths became extinct. Extinctions have also been traced back as far as half a billion years.

A certain periodicity was seen in these extinctions by some investigators of the data. A suggested explanation was the presence of a dark solar companion in a highly elliptical orbit which at perihelion disturbed comets and caused them to fall into the inner solar system, where some collided with Earth and caused the extinction. The hypothetical dark star became known as Nemesis. The fact that the extinctions did not occur quickly but seemed to have been spread out over hundreds if not thousands of years cast doubts on a single-impact hypothesis but supported the viewpoint that a dark star could disturb comets for hundreds of years and it might then take thousands of years for sufficient impacts to occur for many species to be completely eliminated. Another possibility for a periodic disturbance of comets may be a cyclic passage of the Sun and its planets through the plane of the Galaxy and encounters with molecular interstellar clouds which are concentrated in that plane.

Extinctions might also have been caused by impacts on the Moon as well as on Earth. Such impacts would push debris into space from where it could enter orbit around Earth to give rise to transient sunlight-shielding rings. Evidence for such rings is possibly that of the tektite fields which have been discovered stretching long distances across the surface of our planet.

An explanation of the mass extinctions might be that a tenth planet rather than a dark star disturbs the comet cloud when at aphelion, by cutting through the cometary disk at a distance of about 9 billion miles (14.5 billion km) from the Sun. The long-term periodicity developed from a 1000-year orbit might result from the tenth planet having displaced many comets already, so that there is no longer a cloud of comets enveloping the Solar System, but rather groups of comets orbiting the system.

Undoubtedly the search for a trans-Neptunian/Plutonian planet will continue and the greatest chance of success would appear to be when the Space Telescope and the high-technology infrared telescope SIRTF can be placed in orbit. However, there will be many other demands on the use of these instruments since the disruption of the space shuttle program has placed most of the U.S. space program into a hold situation from which recovery will take a long time.

Following Voyager 2's exploration of the Neptunian system a new outer Solar System adventure begins. Both Voyager 1 and Voyager 2 are expected to remain active and performing well for many years to come unless there is some unexpected failure. As Voyagers 1 and 2 head out of the Solar System (figure 6.6), they have important tasks to perform in finding out how interstellar space interfaces with the heliosphere, the region of the Sun's influence. They will supplement the work being performed by Pioneers 10 and 11 which are also traveling out in the Solar System. Pioneer 10 crossed the orbit of Neptune in 1983. It was the first spacecraft to travel beyond all the known planets because at that time Pluto was within the orbit of Neptune. These four spacecraft provide a unique opportunity to extend the charting of the Solar System to an unprecedented extent and to start looking out from its borders at the Galaxy itself; for example, gathering new information about ultraviolet emissions from galactic sources. This opportunity is unlikely to be repeated for many decades if current allocation of national priorities is any guide—at least insofar as the United States is concerned.

As long as Congress allocates funds to keep the ground crews operating, these four spacecraft will continue to send priceless information from the outer reaches of the Solar System. Their internal power supplies are expected to last into the next century and as the power supplies degenerate, energy can be conserved within the spacecraft so that data samples will be taken and transmitted to Earth at reduced rates. Ultimately both Voyager and Pioneer spacecraft will escape completely from the Solar System, but communications could be maintained for as long as the spacecraft continue to function.

The primary task of all these spacecraft is to explore to the limits of the heliosphere (figure 6.7) and to find out where the magnetic influence of the Sun ends and interstellar space begins. This boundary is referred to as the heliopause. If all goes as expected, all four spacecraft should be alive and well at the crossing of the heliopause, and the first to do so may be one of the Pioneers. The exact location of the heliopause is unknown but is believed to be closer to the Sun in the direction that the Sun is moving through the Galaxy. Various estimates have been made, ranging from 50 to 150 times Earth's distance from the Sun; that is from 4.6 to 14 billion miles (7.5 to 22.5 billion km). The extent of the heliosphere will also be dependent upon the level of solar activity, being largest when the Sun is most active.

A restriction on the operation of the Voyagers might arise when the spacecraft achieve a distance at which their attitude-controlling Sun sensors no longer receive sufficient light from the Sun to operate properly. This is expected to occur at about 7.5 billion miles, in the first decade of the twenty-first century. The power produced by the thermoelectric generators will also fall toward a minimum operating value about ten years after the Sun sensor can no longer be used. That will be about 11 billion miles (18 billion km) from the Sun.

There is also the important task of measuring the Sun's magnetic field at

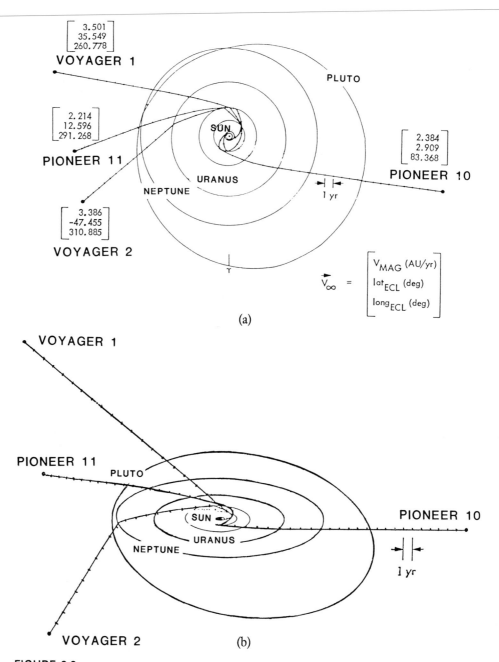

$$\vec{V}_{\infty} = \begin{bmatrix} V_{MAG} \ (AU/yr) \\ lat_{ECL} \ (deg) \\ long_{ECL} \ (deg) \end{bmatrix}$$

(a)

(b)

FIGURE 6.6
Paths of Voyagers and Pioneers out of the Solar System; (a) projected on the ecliptic plane, and (b) an oblique view showing how the Voyagers go above and below the ecliptic plane. (NASA)

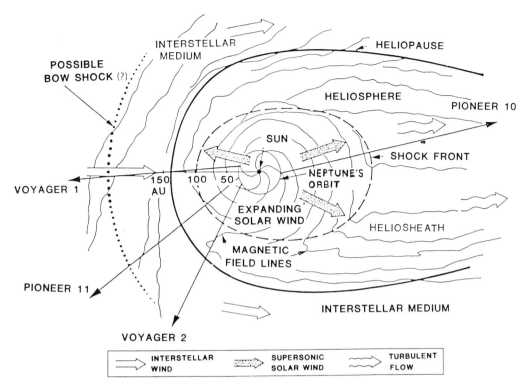

FIGURE 6.7
Diagram of the heliosphere showing its various regions that will hopefully be explored by the Pioneers and Voyagers. Pioneer 11 should encounter the bow shock in the ecliptic plane and the two Voyagers above and below it. Pioneer 10 travels down the magnetic tail of the Solar System (NASA).

great distances and recording the effects of the reversal of this field, which occurs every 22 years. The effects far out in space will be of particular interest since they cannot be observed without spacecraft at the fringes of the Solar System.

An important aspect of these observations is that Voyager 1 is moving high above the plane of the ecliptic so that it can observe the magnetic effects at high Solar System latitudes.

The heliopause, should be affected by the motion of the Sun. At the forward end there would be expected an equivalent of the bow shock observed at planets. This would probably be a region of turbulent motions of charged particles. Voyager 1 is headed in the direction of this leading edge.

The Voyagers and Pioneers will also seek low-energy cosmic rays from the Galaxy, particles which cannot normally penetrate far into the heliosphere. The Pioneers will be in space within the heliosphere to observe and compare data from several solar cycles of activity. The anomalous component of cosmic rays is extremely sensitive to the solar cycle; most particles of the cosmic rays are excluded from the inner Solar System at periods of solar maximum activity. Determining how the cosmic rays enter our Solar System and how they are modulated by the fields generated by the Sun is crucial to designing experiments to seek anti-protons which are of importance not only for studies of atomic physics and physics of our Galaxy, but also in the development of anti-matter drives for interstellar probe missions. Pioneer 10 is expected to reach the modulation boundary for cosmic rays in 1988 at about 42 astronomical units (4 billion miles or 6.4 billion km) from the Sun.

The spacecraft will look for evidence of interstellar hydrogen and helium entering the Solar System from an interstellar wind. At great distances from the Sun the densities of neutral (uncharged) atoms of hydrogen and helium can be ascertained from the intensity of ultraviolet radiation and checks can be made to find out if the heliopause turbulence is stopping hydrogen atoms from flowing into the Solar System from the Galaxy.

Radio emissions from the Sun will be monitored in regions of space where there are virtually no planetary sources to interfere.

Far out in space, beyond the planets and the heliopause, is a region which theory has suggested contains a swarm of cometary bodies known as the Oort Cloud (after the Dutch astronomer, Jan Oort, who first suggested it in 1950). The presence of such a cloud is needed to explain the preponderance of comets whose orbits are close to parabolic and from which the short-period comets eventually evolve. However, there are too many short-period comets to be explained in this way, unless there were more long-period comets in the past than have been observed during the last few hundred years. It has been suggested that the Oort cloud might have a dense inner core to account for the differences in numbers of short- and long-period comets. Such a core would also solve the problem of the Uranus–Neptune perturbations without there having to be a tenth planet.

The nuclei of comets are thought to be captured from interstellar space and literally caused to buzz around the Solar System, held there by the feeble solar gravity at a distance of many billions of miles. While all four spacecraft will eventually pass through this region, they will be inoperable because of loss of power and would also, if still powered, have lost communication with Earth, so we will not be aware of their exploration. This is a pity because confirmation of the distribution of the comets in the Oort cloud would have important implications concerning the comets themselves and concerning the origin of the Solar System and its planets. We might find theoretical connections to the dust clouds recently discovered around other stars at about the same distance from the stars as the Oort cloud is postulated to be from our Sun.

After traversing the region of the Oort cloud, the spacecraft will make their way toward other star systems. Affixed to each spacecraft is a message to extraterrestrials originally suggested by the author for the two Pioneers and implemented by Frank Drake and Carl Sagan. The Pioneers carry small plaques carrying a fundamentally simple message stating from where they came and when, expressed in scientific terms that have a reasonable chance of being interpreted by other intelligent beings. The Voyagers carry much more sophisticated and involved records designed by a team of people. The records encompass some scientific information but also many sociological and political aspects of human life on Earth today, which may not make sense in a cosmic context unless the other beings who find the records have experienced evolution similar enough to terrestrials for them to interpret what the human-related record of political speeches, music, and terrestrial scenes shows.

Because Voyager 2 is expected to fly over the north pole of Neptune to rendezvous with Triton, the spacecraft will afterward continue along a path carrying it far below the ecliptic plane. The two Voyagers thus travel above and below the plane of the ecliptic while the Pioneers keep closer to it. One Pioneer heads in the direction of Gemini, the other toward Sagittarius. The four spacecraft are expected to pass relatively close to other stars in the far distant future. But during the eons of the spacecrafts' journeys the stellar background will change greatly because of the motion of the Sun and the stars around the Galaxy's center. Voyager 1 is expected to pass close to a star in the constellation Camelopardus about 40,000 years hence; Voyager 2 will fly within two light years of a star in Andromeda at about that same time, and over a quarter of a million years later the spacecraft is expected to fly close to the system of Sirius, the Dog Star, and the brightest star visible in Earth's skies today.

The outer Solar System still presents many enigmas and challenges for future missions. Project Galileo, designed to send an atmospheric probe and an orbiter to Jupiter, points the way to further exploration of the large outer planets. Originally scheduled for launching in 1986, it was delayed considerably by the Challenger disaster. Other missions to the outer planets have often been proposed, but currently there are no authorizations to implement these advanced missions.

As detailed in the 1986 report of the President's National Commission on Space, the United States has the knowhow and the technology for far reaching missions and advanced space projects. Unfortunately the nation does not seem inclined to use this unique capability. Meanwhile other nations concentrate their burgeoning space activities on the inner Solar System and on planning manned flight to Mars and the utilization of space resources rather than on in-depth scientific exploration of the remaining frontier of the outer Solar System and beyond.

There are many new programs of scientific research that need implementing; unmanned probes into the atmospheres of the giant satellites Triton and Titan, followed by landers on these mysterious satellites, probes, orbiters, and ring explorers for Saturn, Uranus, and Neptune, a mission to Pluto and its unusual satellite Charon, and spacecraft designed specifically to explore the Oort cloud and further into interstellar space with long-term missions to seek encounters with nearby star systems and the important continuing search for evidence of life elsewhere than on Earth.

Today the continued exploration of the outer planets presents a unique opportunity for the United States. There is, as yet, no serious competition. This facet of exploration requires two capabilities in which American science and technology currently has a lead that is unfortunately diminishing, because time is not on our side. These capabilities are the ability to navigate and command spacecraft over enormous distances, and the ability to return information over these same distances. It is a special moment in history.

Exploration has always changed outlooks. This may be the real reward from space missions. Through exploration people find relief from futile, nonproductive competition and overcrowding. A start at exploring the outer Solar System has been made with the Pioneers and the Voyagers. We can proceed, if we wish, to further exploration with the potential of tremendous benefits for humankind—benefits that are important not only socially and economically but also spiritually in the sense that we may better understand our role in the creative and evolutionary process. We are already into this fantastically stimulating adventure with its promise of future explorations pushing even beyond the Solar System. Only the collapse of society in general into the new barbarisms which seem to be trying to gain footholds in various regions of our planet and a return to new Dark Ages will prevent the unfolding of the space expansion by the United States, the Soviet Union, or a consortium of other nations. At the present time the people of the United States are in a singular position to take the lead and, by reaching for the stars, offer new hopes and peaceful new goals for our species.

INDEX

INDEX